# Finnish-Soviet symposium on non-timber forest resources

25–29 August 1986
University of Jyväskylä
Department of Biology
Jyväskylä, Finland

The first Finnish-Soviet symposium on non-timber forest resources was organized by the Department of Biology of the University of Jyväskylä in Central Finland. The financial support came from the Environmental Research Council of the Academy of Finland.

The symposium was attended by 10 Soviet and 26 Finnish participants. The Soviet participants came from the Karelian ASSR (5), the Estonian SSR (2) and the Latvian, Lithuanian and Russian Federal SSR (1 from each republic). A total of 25 scientific papers, mainly on wild berries and mushrooms, were presented in the symposium. This number includes also the contributions of two Soviet researchers not attending the symposium personally. In addition to this, posters were presented on four topics.

The programme of the symposium was divided in two parts. The first three days consisted of the oral presentations of the scientific papers. The last two days were used for excursions to small and medium sized enterprises, operating as organized buyers or refiners of non-timber forest resources. These were: Tähtisen sieniasema (Padasjoki), a family enterprise operating as an organized buyer of forest mushrooms; Päijät-Hämeen Luontaistuote (Padasjoki), an enterprise practising the drying and selling of herbs, vegetables and berries; a factory owned by Valio (Suonenjoki) and making jams and juices from wild and cultivated berries; the liqueur fabric of Lignell & Piispanen (Kuopio), using wild berries as raw material for its products, and, finally, a family enterprise named Kauppisen Tuote ja Vihannes (Joensuu), an organized buyer of forest mushrooms and berries.

The need for the symposium had been recognized for a long time, and the meeting of Finnish and Soviet researchers proved to be very useful and encouraging. The symposium proceeded in an inspiring atmosphere, and plans were made about continuing the exchange of the knowledge of non-timber forest resources between Soviet and Finnish specialists. These plans are now being realized: the second Soviet-Finnish symposium on this particular topic will be held in August 1988 in Petrozavodsk, Soviet Union.

Acta Bot. Fennica 136:3–5, 1988

# Classification of non-timber resources in the USSR

A. Ph. CHERKASOV

Cherkasov, A. Ph. 1988: Classification of non-timber resources in the USSR. — In: Vänninen, I. & Raatikainen, M. (eds.), Proceedings of the Finnish-Soviet symposium on non-timber forest resources in Jyväskylä, Finland, 25–29 August 1986: 3–5. Acta Bot. Fennica 136. Helsinki. ISSN 0001-5369 ISBN 951-9469-30-3

Non-timber resources are divided into vegetation resources (food, medicinal, honey and apian, technical and forage ones), wildlife resources (vertebrates and invertebrates), environment conservation resources (including climate regulation, water conservation, soil protection and other functions) and social resources (sanitation, recreation). In commercial forests the non-timber resources make up about 50% of the total economic value and in protected forests about 93%.

Key words: Forest by-products, non-timber resources

*A. Ph. Cherkasov, All Union Research Institute of Silviculture and Mechanization of Forestry, Kostroma Research Station, SU-156013 Kostroma, Prospekt Mira 134, USSR*

## THE TERM "NON-TIMBER RESOURCES"

In spite of the fact that the term "non-timber forest resources" has frequently been used, its meaning has no certain definition. For this reason it is difficult to place these resources in the general classification of forest resources and this complicates their evaluation from the forestry, economic and social viewpoints, making it difficult to study and develop them.

The term "non-timber resources" appeared in the USSR about 25 years ago, replacing the term "resources of forest by-products". The latter has been employed up to now in official documents (for instance in the Forestry Code of Russia), but in research and popular papers the term "non-timber resources" has been used increasingly widely. Evidently, it was first used because it sounded better than the rather hackneyed "forest by-products", but later it became clear that the older term does not give a proper idea of the volume and the economic and social importance of non-timber resources.

Forest uses in the USSR are divided into main wood production, in natural forests and secondary wood production from slash and by-products; other forest uses are omitted as being unimportant.

## CLASSIFICATION OF NON-TIMBER FOREST RESOURCES

Non-timber uses of forest include: hay-making, cattle grazing, apiculture, sap and resin collection, gathering fruits, berries, nuts, mushrooms and plants used for medicinal, culinary and decorative purposes, and industrial uses of moss, reeds and forest litter.

Officially, non-timber forest uses include: resin tapping, hunting, fishing and many other special uses. In practice these uses embrace many things that are not associated with wood production and cannot be planned because of their casual nature. Earlier, non-timber forest production was not included in economic surveys, but was treated as worthless and taken for granted.

With the intensification of forest management, non-timber resources are growing in importance and are being included in economic planning. Forest by-products are gradually losing their casual character and becoming special fields of forest management.

Experts estimate that the value of forest by-products equals or exceeds the value of timber. In some experimental forest enterprises, forest by-products are much more profitable than sales of timber. In such cases difficulties arise in defining basic and secondary, principal and by-product resources. Economically, by-products are often the principal resources and in future they can be expected to play the main role in human society and to be rated equally with timber production in all types of forests, including commercial.

When these considerations are taken into account, it is clear that the term "non-timber forest resources" is more objective than "forest by-product resources". We believe that all forest resources should be divided into timber, non-timber and special resources. The use of forest resources includes harvest of timber, stumps, bast, birch, bark, fir, pine and spruce twigs etc. — practically everything that is associated with a tree. We propose to divide non-timber forest uses into:

*Vegetation*
1. Food
   a) Wild fruits (nuts, berries)
   b) Mushrooms
2. Medicinal plants
3. Honey plants
4. Plants for industrial use
5. Forage

*Wildlife*
1. Vertebrates
   a) Game (birds, animals, fish)
2. Invertebrates

*Environment conservation and social uses*
1. Environment conservation
   a) Climate regulation
   b) Water regulation and water conservation
   c) Soil protection
2. Social uses
   a) Sanitation
   b) Recreation.

The two last-named are sometimes called intangible resources.

Non-timber uses include primarily edible plants (wild fruits, nuts, berries, mushrooms), medicinal plants (including poisonous fungi), plants used by bees and in industry (mainly phenol combinations (tannin), dyeing, volatile oil), and forage plants.

The wildlife includes important vertebrates (here mainly game belongs: birds, animals, fish) and invertebrates that play a significant role in the life of a forest.

Non-productive uses can be subdivided into environment conservation and social uses. The former includes regulation of climate and water, and the water-preserving and soil-protecting functions of the forest. The social uses include sanitation, recreation, defence, etc.

Mosses and reeds are a special group of non-timber vegetation resources, which are harvested in forests and forest areas and are widely used in construction, the paper industry etc.

Oleoresin and sugary sap are resources that are difficult to classify; they are close to wood resources, as they are tree products, but might be referred to non-timber resources. At present resin tapping is considered to be an independent kind of forest use, while sap collection is referred to side-line forest use (Forest Code of the Russian Federation, 1978). All kinds of products grown and harvested in forest areas as a result of agricultural activity (cereals, forage crops, etc.), horticulture, cattle breeding, fish production in forest ponds etc. should be referred to a special category, for instance, specialized management, though they may develop from non-timber resources (e.g. berry plantations).

## ECONOMIC VALUE OF NON-TIMBER FOREST RESOURCES

Let us now analyse the economic value of some non-timber uses of forest. There are about 100 types of wild fruit, berry, and nut plants in the territory of the USSR. They cover about 10 million ha, including over 8.5 million ha of berry stands, and produce millions of tons of fruits, berries and nuts (the annual harvest is a little over 100 000 t). In addition, many types of edible plants grow in forests, such as *Heracleum* spp. (cow parsnip), *Urtica* spp. (nettle), *Rumex* spp. (dock), *Aegopodium podagraria* (goat weed), *Pteridium aquilinum, Allium victorialis*, etc.

About 200 of the macromycete species are edible. They can be found in areas ranging from steppe birch woods to tundra forest and their annual yield reaches 5–6 million t.

There are more than 2 500 medicinal plants in the USSR, and over 600 species may be used in the chemical and pharmaceutical industries. Most of these plants grow in the forest zone. The annual industrial demand in the USSR for plant raw material is 60 000 t and 30 000 t are harvested in wild thickets and forest. The forest contains a great many melliferous herbs; there are more than 2 million ha of stands in which lime tree (*Tilia* spp.) are dominant and over 50 million ha of clear-cut and burnt-over areas in which such valuable melliferous plants grow as pea-tree (*Caragana*), raspberry (*Rubus idaeus*), willow-herb (*Epilobium*), willow (*Salix*), *Viburnum*, heather (*Calluna*), mountain ash (*Sorbus*) and many others.

A lime-tree stand of 1 ha can produce about 500 kg of honey, 1 ha of willow-herb about 350 kg of honey, and 1 ha of willow up to 150 kg of honey. It is estimated that our forests can produce about 400 000–500 000 kg of honey annually.

Areas of the state forest fund produce large amounts of grass for cattle, from 300 to 9 000 kg/ha depending on the conditions. Forest rangelands cover 60 million ha in the state forest fund of the USSR and over 300 000 t of hay are harvested annually. In summer 45 million ha of forest are used for grazing about 20 million head of livestock. If we disregard the negative consequences of cattle grazing in forest, and take into account the purchase price of the products of cattle-breeding obtained by grazing in the forest fund, we must conclude that in a number of forest regions cattle-grazing is more profitable than the stumpage price.

Many animals feed on the fruits of forest plants: acorns (*Quercus* spp.), chestnuts (*Castanea* spp.), berries of the mountain ash (*Sorbus* spp.), black elder (*Sambucus* spp.), sweetbrier (*Rosa* spp.), *Viburnum*, honeysuckle (*Lonicera* spp.), raspberry (*Rubus* spp.), bilberry (*Vaccinium myrtillus*), cranberry (*Oxycoccus palustris*) and many other plants. Animals (squirrel, deer, moose, wild boar, mouse-like rodents and others) also feed on many types of mushrooms.

Wildlife resources are abundant in forest: the wildlife and game comprise scores of species. The forests of the state forest fund make up about 45% of the total hunting areas in the USSR. Hunting gives an income equal to one-third of the stumpage price. The greatest populations of the squirrel and sable inhabit the dark coniferous taiga, which covers 147 million ha in the USSR. The moose population in the USSR number 820 000, the average weight of a moose being 400–450 kg. Annually, 200 000 moose can be hunted for meat production, which makes about 100 000 t in dead weight. The wild boar population in the USSR forests is 350 000 and if one-third of this population is culled annually it will provide about 20 000 t of meat.

The skins of some game animals are highly valuable: bear (*Ursus arctos*), beaver (*Castor fiber*), squirrel (*Sciurus vulgaris*), marten (*Martes martes*), fox (*Vulpes vulpes*), wolf (*Canis lupus*), etc. Sable fur is called soft gold. Valuable game birds are: wood grouse (*Tetrao urogallus*), black cock (*Tetrastes bonasia*), wood cock (*Lyrurus tetrix*) and others.

Forest adders (*Vipera berus*) give valuable venom, which is superior to the venom of the cobra and gurza snake. Ants and soil animals also play an important role in forest life.

The forests of the USSR contain over 18 million ha of lakes and streams populated by various species of fish. Every ha of water produces 50 kg of fish, which makes a total of 900 000 t.

Non-timber resources include so-called intangible forest benefits. Efforts have been made lately to assess their economic value, but this will always be difficult. It is clearly impossible, however, to overestimate these benefits. Forest functions such as climate regulation, water conservation and soil protection are invaluable. The use of forest for recreation is also increasing rapidly.

In summing up the value of different groups of non-timber uses, it should be stressed that the total value of non-timber uses in commercial forests makes up about 50% of the total economic effect and in protected forests 93%. Clearly, the forest resources are mainly non-timber products.

Experts in the non-timber line must direct their efforts to active and effective study of the separate components of non-timber uses, and of the interaction of these components, to gain a complete picture of the non-timber forest resources. Much attention should be given to working out methods and ways of evaluating such resources. Without these, the establishment of non-timber uses will be difficult from both the scientific and practical point of view. Greater attention should be given to studies of non-timber uses that are changed by forestry practices (felling, thinning, melioration, reforestation, etc.). These studies will help to give an objective and realistic idea of the forest as a complex natural phenomenon and must be carried out in the immediate future.

I hope that the classification of non-timber uses of the forest given here will help to promote such studies, but I do not imagine that it gives a complete picture of the non-timber potential and I will be quite satisfied if some elements of it are used for making a more effective and complete scheme.

Received 28.VII.1987

Acta Bot. Fennica 136:7–8, 1988

# A model of rational non-ligneous forest resource consumption in intensive forestry in Lithuanian SSR

D. BUDRIUNIENE

Budriuniene, D. 1988: A model of rational non-ligneous forest resource consumption in intensive forestry in Lithuanian SSR. — In: Vänninen, I. & Raatikainen, M. (eds.), Proceedings of the Finnish-Soviet symposium on non-timber forest resources in Jyväskylä, Finland, 25–29 August 1986: 7–8. Acta Bot. Fennica 136. Helsinki. ISSN 0001-5369 ISBN 951-9469-30-3

In intensive forestry, the rational consumption of non-ligneous vegetation having constant area and stable or increasing productivity is possible only in specially delimited areas. The first stage of the model implementation comprises evaluation of non-ligneous forest resources and identification and separation of areas for specialized management. These operations are undertaken by the forest management enterprise "Lesproyekt". The next stage is the application of special forestry arrangements in areas with valuable non-ligneous vegetation. In the Lithuanian Scientific Institute of Forestry, methods have been developed for evaluation of berry and herb resources and delimitation of suitable areas. Measures for the protection and increase of the productivity of the berries of *Arctostaphylos uva-ursi* (L.) Spreng. and *Vaccinium vitis-idaea* L. are already being implemented.

Key words: Consumption, model, non-ligneous forest resources

*D. Budriuniene, Lithuanian Research Institute of Forestry, SU-234321 Kaunas-Girionis, Lithuanian SSR, USSR*

Natural forest is a multifunctional system, offering to the human population a wide diversity of products and properties. The attitudes to the production of the non-ligneous forest vegetation — berries, mushrooms, herbs — have varied widely in the course of human history. In recent years this production has undergone a reappraisal in the forests of medium latitudes and the demands for forest-yielded foodstuffs are constantly increasing, in both the Soviet Union and other countries. Places abunding in mushrooms and berries are widely used for recreation and for gathering these forest products as well.

Under intensive forestry conditions, human activities exert a strong impact upon the forest biocoenoses, and these are sometimes radically changed. The possibility of the natural renewal of many plant resources is reduced to a minimum by cultivation of forest stands of high density and productivity, melioration of swamps, ploughing of forest ground before planting, and use of land for recreation. Consequently, some plant species are now listed as rare and close to extermination. Other raw materials abundant in the past, have now become insufficient. It is felt necessary to introduce a system for regulating forest exploitation, including non-ligneous resources.

Not all the effects of intensive forest husbandry are economically desirable. The impact on the non-ligneous vegetation is very strong and can reduce the stands of some useful species. We must note here that the productions of ligneous and non-ligneous plants to some extent compete with each other and maximization of the former leads to reduction of the latter. Reduction of the non-ligneous plant resources takes place in regions of intensive silviculture, where timber production is the main object. Stable, non-ligneous vegetation, with stable or increasing productivity is possible only when the silviculture takes account of or is focused on the non-ligneous vegetation. Specialization in forest husbandry is necessitated by the fact that 1) complex forest output cannot be maximized simultaneously in one and the same forest; 2) demands for multifunctional forest use have increased; 3) the use of natural resources must be planned on a regional basis; 4) the choice of the best form for exploitation of limited resources is necessary. A model for specialized forest husbandry has been constructed in the Lithuanian Scientific Research Institute of Forestry. Forest areas for specialized production are being delimited, including areas for non-ligneous production. The latter areas are intended to constitute 6% of the state-owned forests in the Lithuanian SSR.

In general, we separate three types of non-ligneous vegetation, on the basis of its amount, demand, utilization and natural properties. Type 1 includes those berry-producing plants which need to be protected and increased by semicultural and cultural methods; cranberry (*Oxycoccus* spp.), cowberry (*Vaccinium vitis-idaea*), bearberry (*Arctostaphylos uva-ursi*), in Lithuania. Type 2 has species that must be protected in their most productive natural habitats and used rationally; blueberry (*Vaccinium myrtillus*), bog whortleberry (*V. uliginosum*), lilly-of-the-valley (*Convallaria majalis*), mushrooms of the first category of edibility. Type 3 has species that are not sufficiently used and that are able to regenerate naturally; hazel (*Corylus avellana*), buckthorn (*Crataegus* spp.), raspberry (*Rubus idaeus*).

The first stage of implementation of the model demands the resource to be evaluated qualitatively and quantitatively. In the Lithuanian SSR, the forest management enterprise "Lesproyekt" has been working on such an evaluation of berries and herbs since 1976, following our suggestions. A total of 20 exploitable plant species have been surveyed. For every species two indicators are evaluated visually: frequency (part of a forest stand with the population of the non-ligneous species under evaluation, %) and density (projective cover of the population, %). Office processing of the data yields the projective cover of the species in the forest taxation plot and the relative area of compact density, i.e. transformed to a 100% projective cover, is calculated. Data yields are grouped, with reserve and productive plant areas separated. From the projective cover and the forest stand density, three productivity grades for non-ligneous plant areas are distinguished. The results are presented in the form of descriptions and summarizing tables of the compact density areas of every non-ligneous species for separate administrative units, for forest stand density and age classes, and for different soil types. Plans have been issued with these species and their productivity grades shown in different colours.

These tentative data form the basis for numerous solutions for rational exploitation and protection projects: location of purveying centres, targets for separate administrative regions and forest husbandry enterprises. Summarizing tables of the areas of non-ligneous plant habitats, with details of the predominant tree species and ages, open prospects for predicting their behaviour. The same data are used to separate areas requiring special management modes for protection and extension. Such areas are evaluated on the basis of their non-ligneous plant habitat densities, tree stand density and age classes, and the potential productivity of the non-ligneous plants according to their biocoenoses, their density, and their distance from urban centres and recreation areas.

As the next step of the model implementation, integral forestry areas must be established, which will protect and extend the productivity of specified exploitable plant species. In the Lithuanian Scientific Institute of Forestry, specific regulations for silviculture in areas of productive non-ligneous vegetation are being prepared. Our suggestions for the protection and increased productivity of the bearberry are already being implemented.

In modern forestry, which is able to control the biocoenosis, owing to the advances of science and technology, due attention must be paid to non-ligneous plants. Socialist governments, with their state-owned forests and planned management strategies, have an excellent basis for scientific and rational forestry policies.

Received 28.VII.1987

Acta Bot. Fennica 136:9–10, 1988

# Estimates of wild berry yields in Finland

MIKKO RAATIKAINEN

Raatikainen, M. 1988: Estimates of wild berry yields in Finland. — In: Vänninen, I. & Raatikainen, M. (eds.), Proceedings of the Finnish-Soviet symposium on non-timber forest resources in Jyväskylä, Finland, 25–29 August 1986: 9–10. Acta Bot. Fennica 136. Helsinki. ISSN 0001-5369 ISBN 951-9469-30-3

Thirty-six of the wild plant species growing in Finland have edible berries. The annual mean yields (million kg) are: *Empetrum nigrum* 350, *Vaccinium vitis-idaea* 180–200, *V. myrtillus* 150–200, *V. oxycoccos* + *V. microcarpum* < 25, *Rubus chamaemorus* 25, *Vaccinium uliginosum* 20, *Rubus idaeus* 11, *R. arcticus* < 1, *Fragaria vesca* 1, *Arctostaphylos alpina* 0.1 and other berries 150. The total yield is over 900 million kg.

Key words: *Arctostaphylos alpina, Empetrum nigrum, Rubus, Fragaria vesca, Vaccinium*

*Mikko Raatikainen, Department of Biology, University of Jyväskylä, Seminaarinkatu 15, SF-40100 Jyväskylä, Finland*

Thirty-six of the wild plant species growing in Finland bear edible berries (Raatikainen 1984). Interest in their yields and their harvesting first arose after the fruit failure caused by the severe winter of 1939–1940.

The first estimate of the yield of *Rubus chamaemorus* in Finland was made by Ruusa Merikallio, who assessed the mean yield at 90 million kg. The report of the state fruit committee (Valtion hedelmäkomitean mietintö 1941) presented estimates of the yield of seven wild berry species (Table 1).

Mäkinen (1974) estimated that the total yield of *Rubus chamaemorus* in Finland was 25 million kg. Somewhat later, Veijalainen (1977) compiled all the scattered research results

Table 1. *Estimated mean yields (million kg/year) of wild berries in Finland.*

| | Valtion hedelmä-komitean mietintö 1941 | Veijalainen 1977 | Metsämarja- ja sieni-toimikunnan mietintö 1979 | This survey |
|---|---|---|---|---|
| *Empetrum nigrum* coll. | .. | .. | .. | 350 |
| *Vaccinium vitis-idaea* | 70 | 500[1)] | 200–500 | 200 |
| *V. myrtillus* | 42 | 100 | 100 | 200 |
| *V. oxycoccos* & *V. microcarpum* | 40 | 25–50 | 25–50 | < 25 |
| *V. uliginosum* | .. | .. | .. | 20 |
| *Rubus chamaemorus* | 30 | 25 | > 25[1)] | 25 |
| *R. idaeus* | 10 | .. | .. | 11 |
| *R. arcticus* | 1 | .. | .. | < 1 |
| *Fragaria vesca* | 3 | .. | .. | 1 |
| *Arctostaphylos alpina* | .. | .. | .. | 0.1 |
| Other berries | 4 | .. | .. | 100 |
| Total berry yield | 200 | 650[2)] | .. | > 900 |

1) good berry year
2) yield of four berry plants

and presented yield estimates for four species (Table 1). A few years later, these figures were slightly altered in the report of the commission on forest berries and mushrooms (Metsämarja- ja sienitoimikunnan mietintö 1979) (Table 1).

In the 1970s and 1980s, the berry yields of a number of areas, approximately the size of a commune, were investigated by piecemeal counting (Huttunen 1978, Ruuhijärvi et al. 1978), by the picking method (Raatikainen 1978, Jaakkola 1983, Raatikainen & Raatikainen 1983, Salo 1983, Raatikainen et al. 1984) or by a combination of the two methods (Raatikainen et al. 1985). On the basis of his own investigations, Huttunen (1978) estimated the yield of *Vaccinium oxycoccos* and *V. microcarpum* in the whole country at nearly 50 million kg, and the yield of *Rubus chamaemorus* at approximately 5 million kg. Raatikainen et al. (1984) estimated the size of the *Vaccinium vitis-idaea* yield in Finland at 180–200 million kg and the size of the *V. myrtillus* yield at 150–200 (250) million kg. Kujala et al. (1987) estimated in Lapland the size of the *Empetrum nigrum* yield at 215 million kg and the size of the *Vaccinium uliginosum* yield at 8 million kg.

On the basis of all the above-mentioned and some unpublished data, new estimates have now been prepared of the berry yields in Finland. According to these, the edible berry yield in the whole country averages over 900 million kg annually (Table 1).

The yields of *Vaccinium vitis-idaea* and *V. myrtillus* were underestimated in 1941. It is probable that the yields at that time were roughly the same size as the present-day yields in Sweden (see Eriksson et al. 1979).

The yields of the mire species *Rubus chamaemorus, Vaccinium oxycoccos* and *V. microcarpum* were probably also underestimated in 1941. These yields did not decrease significantly until ditching of the mires, which began after 1950.

As regards the berries growing in meadows, the yields of *Rubus arcticus* and *Fragaria vesca* have decreased greatly since the 1940s, due to changes in their habitats.

Among the highly productive species, the mean yield of *Sorbus aucuparia* is probably greater than the yield of *Rubus chamaemorus*, but smaller than the yield of *Vaccinium myrtillus*. The yield of *Hippophae rhamnoides* is probably smaller than the yield of *Rubus chamaemorus*. Accurate information on the yields of these species should be acquired and more precise data obtained on the yields of the other species.

## REFERENCES


Eriksson, L., Kardell, L. & Ingelög, T. 1979: Blåbär, lingon, hallon. Förekomst och bärproduktion i Sverige 1974–1977. (Summary: Bilberry, lingonberry, raspberry. Occurrence and production in Sweden 1974–1977.) — Sveriges Lantbruksuniversitet. Avdelningen för Landskapsvård. Rapport 16:1–124 + 3 bilagor.

Huttunen, A. 1978: Hilla- ja karpalosadoista Siuruan alueella. (Summary: On the cloudberry and cranberry yields in Siurua district, N.-Finland.) — Suo 29:17–21.

Jaakkola, I. 1983: Rovaniemen maalaiskunnan marjasatoinventointi. — Metsäntutkimuslaitoksen tiedonantoja 105: 137–143.

Kujala, M., Kynsilehto, K., Ohenoja, E., Saastamoinen, O. & Sepponen, P. 1987: Lapin läänin luonnonmarja- ja sienivarat, niiden satoarviot, hyödyntäminen ja sivuansiollinen merkitys. — Pellervo-seuran markkinatutkimuslaitos. Raportti, 60 s.

Mäkinen, Y. 1974: Suomuuraimen viljelystä. (Summary: On the cultivation of cloudberry.) — Suo 25:65–70.

Metsämarja- ja sienitoimikunnan mietintö. — Komiteanmietintö 1979, 19: I–XII, 1–67 + 1–13 Tables + 7 Figs. + 1–4 Appendices.

Raatikainen, M. 1978: Puolukan sato, poiminta ja markkinointi Pihtiputaan kunnassa. (Summary: The berry yield, picking, and marketing of Vaccinium vitis-idaea L. in the commune of Pihtipudas.) — Silva Fennica 12:126–139.

Raatikainen, M. 1984: Syötävien luonnonmarjojemme sadoista ja poiminnasta. (Summary: Yields and harvesting of berries in Finland.) — Academia Scientiarum Fennica, Year Book 1982–1983:135–158.

Raatikainen, M. & Raatikainen, T. 1983: Mustikan sato, poiminta ja markkinointi Pihtiputaalla. (Summary: The berry yield, picking and marketing of Vaccinium myrtillus in the commune of Pihtipudas, northern Central Finland.) — Silva Fennica 17:113–123.

Raatikainen, M., Rossi, E., Huovinen, J., Koskela, M.-L., Niemelä, M. & Raatikainen, T. 1984: Metsä- ja suomarjasadot Väli-Suomessa. (Summary: The yields of the edible wild berries in Central Finland.) — Silva Fennica 18:199–219.

Raatikainen, M., Rossi, E. & Vänninen, I. 1985: Kotipihlajan marjasadon määritysmenetelmät ja marjasato. (Summary: On methods of estimating the berry yield and the berry yield of rowan tree (Sorbus aucuparia).) — Silva Fennica 19:203–209.

Ruuhijärvi, R., Kerkelä, T. & Leivo, A. 1978: Ounasjokitutkimuksia IV. Tepaston ja Meltauksen allasalueiden marjasadosta. — 23 + 2 s. Helsinki.

Salo, K. 1983: Marja- ja sienisatojen seuranta VMI-tutkimuksen osana Pohjois-Karjalassa. — Metsäntutkimuslaitoksen tiedonantoja 90:122–134.

Valtion hedelmäkomitean mietintö. II osa. — Komiteanmietintö Mon. 1941, 6:1–178 + 5 Appendices.

Veijalainen, H. 1977: Luonnonmarjasadot ja niiden käyttö. — Molekyyli 34, 1:5–6.


Received 28.VII.1987

Acta Bot. Fennica 136:11–13, 1988

# Ten years of inquiries on the berry and mushroom yields in Finland, 1977–1986

MATTI KUJALA

Kujala, M. 1988: Ten years of inquiries on the berry and mushroom yields in Finland, 1977–1986. — In: Vänninen, I. & Raatikainen, M. (eds.), Proceedings of the Finnish-Soviet symposium on non-timber forest resources in Jyväskylä, Finland, 25–29 August 1986: 11–13. Acta Bot. Fennica 136. Helsinki. ISSN 0001-5369 ISBN 951-9469-30-3

The Marketing Research Institute of the Pellervo Society produces estimates and forecasts of the wild berry and mushroom yields in Finland since 1977. The so-called Marsi inquiry produces also estimates on the incomes of the country people from berry and mushroom picking. The inquiry is done by 200 local reporters in 18 districts of the country. The most important commercial berries are the lingonberry, the bilberry and the cloudberry. The interest in the mushroom trade has increased in recent years. The tax-free incomes of the country people from the wild berries and mushrooms have been 35–93 million marks annually.

Key words: Berry trade, Marsi inquiry, mushroom trade

*Matti Kujala, Marketing Institute of the Pellervo Society, P.O. Box 77, SF-00101 Helsinki*

In Finland country-wide estimates and forecasts on the wild berry and mushroom yields have been made annually from 1977 onwards by the Marketing Research Institute of the Pellervo Society. This generally useful survey is financed by the Ministry of Agriculture and Forestry and supported by the organized trade and processing industries in the field of berry and mushroom marketing.

The so-called Marsi inquiry produces three estimates of the wild berry and mushroom yields during the harvesting season in July, August and September. At the end of the season, an additional inquiry is made in order to collect information about the quantities of berries and mushrooms bought at the collecting points by the merchants in the field. At the same time, the prices paid to the pickers are asked, so that it is possible to calculate the incomes from berry and mushroom picking by districts and in the whole country.

The Marsi inquiry is based on the work of 200 local reporters during the season. They are asked to make observations in at least 10 different berry and mushroom forests within the commune in which they live. Finland is divided into 18 districts for this purpose, and each district has its own district reporter, who is assisted by the local observers. The district reporter reports the yield profile to the head-office of the Marketing Research Institute of the Pellervo Society, where the estimate of the yields is produced immediately with the help of a computor program and distributed to the buyers and pickers of the berries and mushrooms concerned via the mass media.

Every year after the buying season the information about the quantities of berries and mushrooms received by the organized buyers is used to compile general statistics on the berry and mushroom trade in the country. By asking the picker-prices, it is possible to calculate the weight prices and total incomes can be followed. In addition to the quantities picked for sale the amounts of berries and mushrooms picked by households for their own use have been estimated to be at least 10 times the quantities sold.

For decades the most important commercial crop has been the lingonberry, but during recent years the bilberry has become more significant. The cloudberry is a very valuable source of income especially for the local people in northern Finland, because its price has always been very high, due to its rarity and the variations in the yields.

During the last decade, the quantities of lingonberries bought by the organized merchants varied from 10.2 million kg to 1.7 million kg, according to the yield and the marketing situation. During the same period, the purchases of bilberries have been between 5.2 million kg and 0.7 million kg per year and those of cloudberries between 1,1 million kg and 0.07 million kg (Fig. 1).

The commercial value of mushrooms is not yet as significant as that of berries, but the interest in the mushroom trade has increased in recent years (Fig. 2). Chantarelles (*Cantharellus cibarius*) are picked so much by the individual households that the organized merchants are not able to buy them in sufficient quantities. During the last 10 years, the best harvest has been only 78 000 kg. Purchases of *Boletus* species have

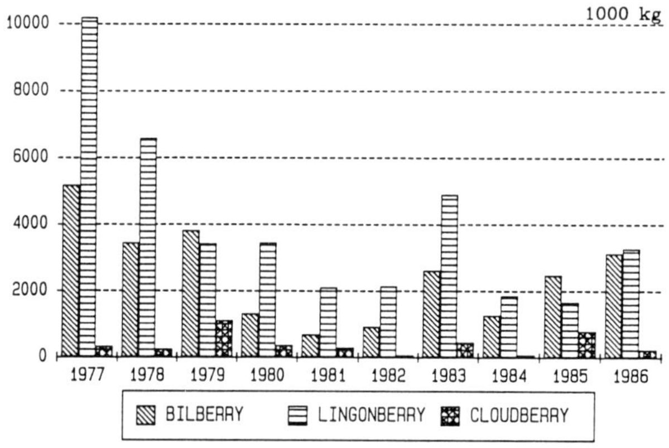


*Fig. 1.* PSM Marsi/quantities of wild berries received by the organized merchants in 1977–1986.

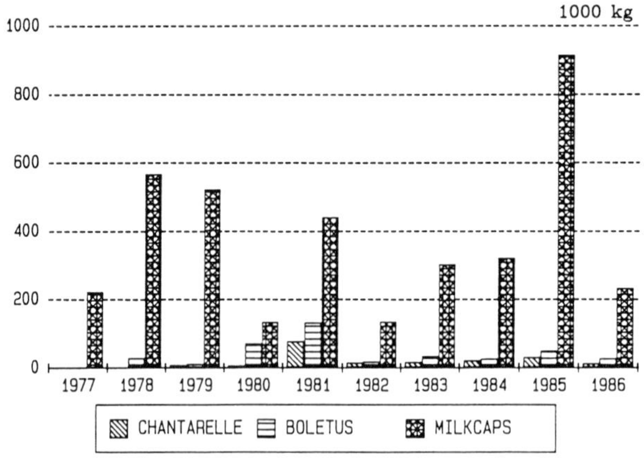


*Fig. 2.* PSM Marsi/quantities of wild mushrooms received by the organized merchants in 1977–1986.

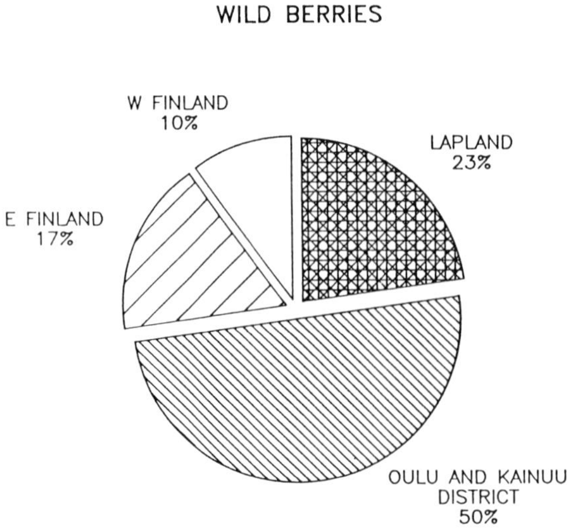


*Fig. 3.* PSM Marsi/distribution of picking incomes in Finland in 1986.

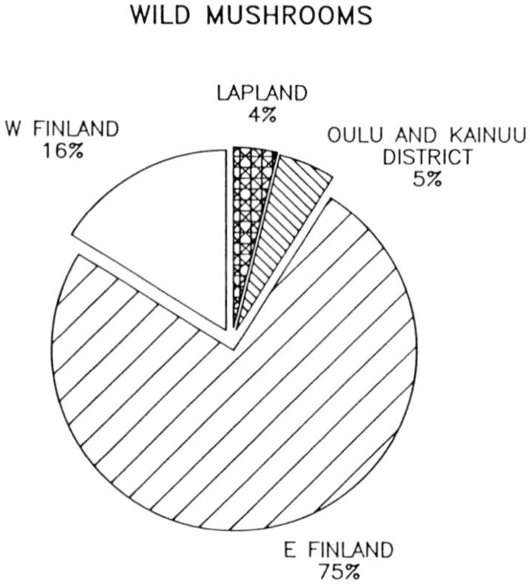


*Fig. 4.* PSM Marsi/distribution of picking incomes in Finland in 1986.

Table 1. *Incomes in FIM from picking wild berries and mushrooms for sale.*

| Year | FIM |
|---|---|
| 1977 | 73 014 000 |
| 1978 | 48 398 000 |
| 1979 | 51 791 000 |
| 1980 | 59 670 000 |
| 1981 | 39 233 000 |
| 1982 | 55 689 000 |
| 1983 | 93 257 000 |
| 1984 | 34 941 000 |
| 1985 | 63 495 000 |
| 1986 | 49 066 000 |

amounted to 132 600 kg when the yield has been very abundant. The most important mushrooms for domestic consumption are milk caps (*Lactarius* spp.). These are picked by the households every year and also received by the merchants, especially in eastern Finland. The greatest harvest received for sale during the 10 year period was 915 700 kg in 1985.

Wild berries and mushrooms are mainly picked for sale by the people in northern and eastern Finland. In this way they have earned tax-free incomes of about 35–93 million mk annually (Table 1). These incomes from picking have been distributed between the districts as shown in Figs. 3 and 4. The people in northern Finland are especially interested in picking berries for sale and mushroom picking is very traditional in the eastern part of the country. In these areas wild berries and mushrooms are a significant seasonal source of income to the country people.

Received 28.VII.1987

Acta Bot. Fennica 136:15–18, 1988

# Training people to collect and sell natural products in Finland

MARJA HÄRKÖNEN

Härkönen, M. 1988: Training people to collect and sell natural products in Finland. — In: Vänninen, I. & Raatikainen, M. (eds.), Proceedings of the Finnish-Soviet symposium on timber forest resources in Jyväskylä, Finland, 25–29 August 1986: 15–18. Acta Bot. Fennica 136. Helsinki. ISSN 0001-5369 ISBN 951-9469-30-3

Training of mushroom advisers and pickers was started in Finland in 1969. Education of wild herb advisers was begun in 1984 and courses on berries in 1987. The National Board of Vocational Education takes care of the training.

Key words: Commercial mushrooms, mushroom advisers, natural products

*Marja Härkönen, National Board of Vocational Education, Hakaniemenkatu 2, SF-00530 Helsinki, Finland; Present address: Department of Botany, University of Helsinki, Unioninkatu 44, SF-00170 Helsinki, Finland*

Courses on the collection and use of natural products are held by several organizations and private persons in Finland, but this article describes only an education program in the collection and sale of mushrooms, herbs and berries financed by the Finnish government and arranged by the National Board of Vocational Education.

## MUSHROOMS

Finnish people traditionally pick and use wild mushrooms, but the number of species gathered is rather small. By far the most frequently collected are *Lactarius* species. *Gyromitra esculenta* is also highly appreciated and so is *Cantharellus cibarius* in southern Finland. More enterprising pickers gather some *Boletus* species, *Cantharellus tubaeformis* and *Albatrellus ovinus*. The ways of preserving the mushrooms are also limited, the most usual being salting (Jäppinen et al. 1985, Kujala et al. 1985).

Finland possesses about 100 good, edible, easy-to-use mushroom species and in an average year their yield is about 1 500 million kg. Although 250–400 million kg could easily be reached, only about 1% of the yield is actually collected (Rautavaara 1947, Saastamoinen 1983, Ohenoja 1984, Ohenoja & Koistinen 1984, Salo 1985).

Recently, the consumption of mushrooms has increased in both private homes and catering kitchens. The import of mushrooms, mostly canned champignons, has also grown, but the export has not developed (Jäppinen et al. 1985).

### The training of mushrooms advisers and pickers organized by the National Board of Vocational Education

In order to utilize better the rich mushroom yield of the Finnish forests, training of mushroom advisers and pickers was started in 1969 at the Department of Forestry Education of the National Board of Forestry. In 1973 this department was transferred to the National Board of Vocational Education, which nowadays takes care of the training of mushroom advisers. The aim of the education is to provide every commune in Finland with one or more mushroom advisers. So far 1978 advisers have been trained, but there are still about 70 communes without any mushroom adviser.

### The courses for mushroom advisers

The courses for mushroom advisers are held in 5–7 forestry schools, situated in different parts of the country. At first a candidate must complete a one-week basic course. The instruction and stay in the boarding school are free for the candidate. The teachers are so-called mushroom inspectors, well-known mushroom experts, who keep abreast of the latest mushroom research. Most of them have their main occupation at universities.

After the course, the participants receive a commercial mushroom adviser's certificate, which gives them the right to train commercial mushroom pickers in their home district. They are entered in a register of mushroom advisers kept by the National Board of Vocational Education and anyone who needs the services of a mushroom adviser may consult the register to find the name and address of the nearest adviser.

There are several types of additional courses, where the mushroom advisers can increase their skills year after year. At one of them they are taught how to run small local mushroom stations, which buy and preprocess commercial mushrooms.

The courses are advertised in local newspapers. As they have been very popular, some selection of the candidates is necessary. No special education or knowledge of the subject is demanded. The most important criterion for selection is that the candidate will be willing and able to train pickers and to spread information about natural products.

### Requirements of a mushroom adviser

The mushroom advisers train commercial mushroom pickers in one-day courses. The pickers are taught to identify 1–3 mushroom species properly and to collect and prepare them for sale. After the course the picker receives a commercial mushroom picker's card, which is a guarantee of the quality of the mushrooms for shops, restaurants, market sellers, etc. This card is not compulsory in the mushroom trade in Finland, but many of the larger stores require the pickers to have it.

The mushroom advisers also give general information on mushrooms and arrange mushroom excursions and exhibitions. Most of them work for various advisory organizations, such as Housewives' Associations or Agricultural Associations.

There are so many species of edible mushrooms that it would be an overwhelming task for a person who is not a mycologist to learn all of them. For this reason, mushroom advisers are taught to recognize a limited number of mushroom species, but they are taught to know these thoroughly. In the courses the pupils collect mushrooms daily and arrange an exhibition of them. Special attention is paid to poisonous species and to species that resemble edible mushrooms. The cleaning and preserving of mushrooms is also taught. Only fresh and healthy mushrooms without larvae are accepted for human consumption. The pupils also learn how to divide the commercial mushrooms into three classes according to their shape and size (extra, class I, class II).

The quality, handling and selling of mushrooms is controlled by several regulations in Finland (Härkönen & Mäki 1983). The most important of them are the statute regarding edible mushrooms (Asetus ruokasienistä 1981, Matsvampsförordning 1981) and the decree of the Commerce and Industry Board concerning commercial mushrooms (Elinkeinohallituksen päätös kauppasienistä 1983, Näringsstyrelsens beslut om handelssvampar 1983). Only the so-called commercial mushrooms listed by the Commerce and Industry Board may be sold. The list consists of 15 wild mushroom species or species groups and five cultivated or imported mushrooms. The wild ones are: *Boletus edulis* (including *B. pinophilus* and *B. reticulatus*), *Suillus variegatus*, *Leccinum versipelle* (including *L. vulpinum* and *L. aurantiacum*), *Lactarius trivialis*, *L. rufus*, *L. torminosus*, *L. deterrimus* (including *L. deliciosus*), *Russula paludosa*, *Cantharellus cibarius*, *C. tubaeformis* (including *C. lutescens*), *Craterellus cornucopioides*, *Albatrellus ovinus*, *Hydnum repandum* (including *H. rufescens*), *Gyromitra esculenta* and *Morchella* spp. The cultivated ones are: *Lentinus edodes*, *Agaricus* spp., *Pleurotus ostreatus*, *Flammulina velutipes* and *Tuber* spp.

The mushroom advisers are taught to recognize the wild commercial mushrooms and the following other good mushrooms for domestic consumption: *Suillus luteus*, *Russula decolorans*, *R. obscura*, *R. claroflava*, *Hygrophorus camarophyllus*, *Rozites caperatus*, *Armillaria mellea* coll., *Macrolepiota procera*, *Lepista nebularis*, *L. nuda* and *Hypholoma capnoides*.

The mushroom advisers also have to be able to identify some poisonous fungi. The following species are learned at the courses: *Amanita virosa*, *A. phalloides*, *A. citrina*, *A. muscaria*, *A. regalis*, *A. pantherina*, *A. porphyria*, *Cortinarius speciosissimus*, *C. limonius*, *C. gentilis*, *Entoloma lividum*, *E. vernum*, *Inocybe* spp., *Clitocybe dealbata*, *Coprinus atramentarius*, *Hypholoma fasciculare*, *Paxillus involutus*, *Tricholoma pessundatum*, *Gyromitra esculenta*, *G. ambigua* and *Galerina marginata*.

The knowledge of the mushroom adviser candidates is tested at an examination in which they have to identify the commercial mushrooms among an assortment of living specimens. The assortment always includes species other than the commercial ones, especially those that resemble them.

## WILD HERBS AND PLANTS USED FOR MEDICINAL PURPOSES

The interest in natural herbs has increased in Finland recently (Härkönen et al. 1984) and the sales of health foods have grown from year to year. In most cases the herbs sold in health shops have been imported. Even dried birch and bilberry leaves and heather flowers may originate from Central Europe, although plenty of them are available in the relatively unpolluted Finnish countryside.

The health shop customers often ask for native products. The dealers usually answer that they are willing to sell domestic products, but these are generally not available and are too expensive and of uncertain quality.

### The problems involved

The collection of wild herbs involves more problems than that of mushrooms.

In Finland the so-called everyman's rights allow anyone to collect wild berries and mushrooms on private land, but they do not apply to herbs. From the point of view of nature protection, more caution is needed in picking herbs than in mushroom or berry picking, because the intensive collection of whole plants may threaten their existence in a certain habitat. Therefore only common and abundant plant species should be collected commercially.

There is no law concerning wild herbs like the one for mushrooms. The herb trade is regulated by orders scattered in various laws for food, health care and marketing.

The picking of wild herbs in Finland has been disorganized and since it has been performed by hand, the native products have been rather expensive. Birch leaves, for instance, should be collected in connection with forest cutting, when abundant branches are lying on the ground. In the spring, sap should be taken from birches that are to be felled during the next few years.

### Start of training of wild herb advisers

The National Board of Vocational Education has begun to educate wild herb advisers to identify, collect and process wild herbs and to teach these skills to wild herb pickers. The program does not include the production of cultivated herbs. In spite of many problems, for instance, the lack of machines for intensive collecting, the education of wild herb advisers was begun in 1984 in forest schools, with a program similar to the one for mushroom advisers. The following medicinal herbs and other plants are covered during the courses: *Cetraria islandica* (lichen), *Equisetum arvense*, *Juniperus communis*, *Pinus sylvestris*, *Urtica dioica*, *Betula pendula*, *B. pubescens*, *Vaccinium vitis-idaea*, *Arctostaphylos uva-ursi*, *Vaccinium myrtillus*, *Ledum palustre*, *Calluna vulgaris*, *Ribes nigrum*, *Fragaria vesca*, *Filipendula ulmaria*, *Rubus arcticus*, *R. idaeus*, *Alchemilla* spp., *Sorbus aucuparia, Trifolium pratense, T. repens, Epilobium angustifolium, Aegopodium podagraria, Frangula alnus, Solidago virgaurea, Matricaria recutita, Taraxacum* spp. and *Achillea millefolium.*

Special attention is paid to the following wild poisonous plants: *Polygonatum odoratum, Chelidonium majus, Convallaria majalis, Viburnum opulus, Trollius europaeus, Lonicera xylosteum, Ranunculus* spp., *Actaea spicata, Conium maculatum, Cicuta virosa, Daphne mezereum, Maianthemum bifolium, Frangula alnus, Solanum dulcamara, Caltha palustris, Paris quadrifolia, Ledum palustre, Sambucus racemosa, Equisetum palustre, Calla palustris, Anemone* spp. and *Hepatica nobilis.*

## BERRIES

Wild berries are traditionally picked in Finland for domestic use and export. It was thought that this tradition was adequate and no education was needed. Investigations (Raatikainen 1978, Raatikainen & Raatikainen 1983, Raatikainen et al. 1984, Salo 1985), have shown, however, that much is to be learned regarding the habitats for berries and the number of edible species. Also, the quality of berries brought to markets is not always satisfactory. Unskilled use of a picking instrument, in particular, may spoil their quality, and they may include unripe or spoiled berries or even litter. In the years 1979–1983, the export of Finnish berries diminished from 5 753 to 1 776 t. At the same time the export of Swedish berries diminished from 8 490 to 7 504 t (Projekt för utvecklande av gemensam standard för nordiska skogsbär 1986). Therefore a Scandinavian committee was established in 1985, and its first report concerning the quality standards for Scandinavian wild berries has just appeared (Projekt för utvecklande av gemensam standard för nordiska skogsbär 1986).

To improve the availability and quality of Finnish wild berries, the National Board of Vocational Education has added courses on berries to its program for nature product advisers. The first courses on berries was held in the summer of 1987. The following wild berries are covered: *Juniperus communis, Arctostaphylos alpina, Vaccinium oxycoccos, V. vitis-idaea, V. uliginosum, V. myrtillus, Empetrum nigrum, Rubus chamaemorus, R. arcticus, R. saxatilis, R. idaeus, Fragaria vesca, Sorbus aucuparia* and *Hippophaë rhamnoides.*

Special attention is paid to the following poisonous berries: *Actaea spicata, Daphne mezereum, Frangula alnus, Sambucus racemosa, Lonicera xylosteum, Solanum dulcamara, Convallaria majalis, Polygonatum odoratum, Paris quadrifolia* and *Maianthemum bifolium.*

## TEACHING MATERIAL

As there were no books on the market which would be suitable as textbooks for the courses, the National Board of Vocational Education has published books of its own (Ammattikasvatushallitus 1981, 1983, 1984, 1985, 1987, Statens tryckericentral 1982, 1984, 1985, 1987). One deals with commercial mushrooms, one gives general information and advice about mushrooms, another tells about poisonous mushrooms and mushroom poisoning. One book deals with wild herbs, their collection, use and preservation, and includes general information about the substances contained by herbs and poisonous plants, and the regulations concerning herb picking. A book on wild berries was published in summer 1987.

## REFERENCES

Ammattikasvatushallitus 1981: Myrkkysieniopas. — Valtion painatuskeskus. 31 pp. Helsinki.

Ammattikasvatushallitus 1983: Kauppasieniopas. — Valtion painatuskeskus. 65 pp. Helsinki.

Ammattikasvatushallitus 1984: Sienineuvonnan opas. — Valtion painatuskeskus. 68 pp. Helsinki.

Ammattikasvatushallitus 1985: Kauppayrttiopas. — Valtion painatuskeskus. 72 pp. Helsinki.

Ammattikasvatushallitus 1987: Luonnonmarjaopas. — Valtion painatuskeskus. 58 pp. Helsinki.

Asetus ruokasienistä (871/81, 11.12.1981).

Elinkeinohallituksen päätös kauppasienistä (28.6.1983).

Härkönen, M. & Mäki 1983: Ruokasieniasetus ja kauppasieniluettelo. — Metsäntutkimuslaitoksen tiedonantoja 91: 66–71.

Härkönen, M., Hakala, P., Johansson, T., Raukko, U., Suojanen, A. & Vauras, J. 1984: Yrttineuvojan koulutusta pohtineen työryhmän muistio. — Ammattikasvatushallitus. 22 pp. Helsinki.

Jäppinen, J.-P., Kirsi, M. & Salo, K. 1985: Luonnonvaraisten sienten sadot ja kaupallinen poiminta Itä-Suomessa, ensisijaisesti Pohjois-Karjalan läänissä. — Metsäntutkimuslaitoksen tiedonantoja 200:1–103.

Kujala, M., Ukkonen, E., Nieminen, A. & Meretniemi, I. 1985: Marjojen ja sienten satoarvioista ja kauppaantulomääristä vuosina 1977–1984. Pellervo-Seuran markkinatutkimuslaitos. 33 pp. Helsinki.

Matsvampsförordning (871/81, 11.12.1981).

Näringsstyrelsens beslut om handelssvampar (28.6.1983).

Ohenoja, E. 1984: Fruit body production of larger fungi in Finland. 1. Introduction to the study in 1976–1978. — Ann. Bot. Fennici 21:349–355.

Ohenoja, E. & Koistinen, R. 1984: Fruit body production of larger fungi in Finland. 2. Edible fungi in northern Finland 1976–1978. — Ann. Bot. Fennici 21:357–366.

Projekt för utvecklande av gemensam standard för nordiska skogsbär 1986: Slutrapport. — Pellervo-Seuran markkinatutkimuslaitos. 21 pp. + 80 pp. appendices. Helsinki.

Raatikainen, M. 1978: Puolukan sato, poiminta ja markkinointi Pihtiputaan kunnassa (Summary: The berry yield, picking and marketing of Vaccinium vitis-idaea L. in the commune Pihtipudas.) — Silva Fennica 12: 126–139.

Raatikainen, M. & Raatikainen, T. 1983: Mustikan sato, poiminta ja markkinointi Pihtiputaalla. (Summary: The berry yield, picking and marketing of Vaccinium myrtillus in the commune of Pihtipudas, Northern Central Finland.) — Silva Fennica 17:113–123.

Raatikainen, M., Rossi, E., Huovinen, J., Koskela, M.-L., Niemelä, M. & Raatikainen, T. 1984: Metsä- ja suomarjasadot Väli-Suomessa. (Summary: The yields of the edible wild berries in Central Finland). — Silva Fennica 18:199–219.

Rautavaara, T. 1947: Suomen sienisato. (Summary: Studies on the mushroom crop in Finland and its utilization). — 534 pp. Porvoo & Helsinki.

Saastamoinen, O. 1983: Marjojen ja sienten talteenoton taloudesta. — Metsäntutkimuslaitoksen tiedonantoja 91:41–53.

Salo, K. 1985: Luonnonmarjojen ja sienten poiminta Suomussalmella ja eräissä Pohjois-Karjalan kunnissa. (Summary: Wild-berry and edible-mushroom picking in Suomussalmi and in some North Karelian communes, Eastern Finland.) — Folia Forestalia 621:1–30.

Yrkesutbildningsstyrelsen 1982: Giftsvampar i bild. — Statens tryckericentral. 31 pp. Helsingfors.

Yrkesutbildningsstyrelsen 1984: Handelssvampar i bild. — Statens tryckericentral. 65 pp. Helsingfors.

Yrkesutbildningsstyrelsen 1985: Handbok för svampkonsulenter. — Statens tryckericentral. 69 pp. Helsingfors.

Yrkesutbildningsstyrelsen 1987: Handelsörter i bild. — Statens tryckericentral. 72 pp. Helsingfors.

Received 18.V.1987

Acta Bot. Fennica 136:19–21, 1988

# Yield forecasting and optimization of berry harvesting in the forests of Southern Karelia, USSR

T. V. BELONOGOVA

Belonogova, T.V. 1988: Yield forecasting and optimization of berry harvesting in the forests of Southern Karelia, USSR. — In: Vänninen, I. & Raatikainen, M. (eds.), Proceedings of the Finnish-Soviet symposium on timber forest resources in Jyväskylä, Finland, 25–29 August 1986: 19–21. Acta Bot. Fennica 136. Helsinki. ISSN 0001-5369 ISBN 951-9469-30-3

Nearly 20 species of berry plants can be found in Karelia. Some of them, especially the bilberry and cowberry, have considerable stocks and are widely used. According to the available data, the yield of the bilberry is 100 000 t, and that of the cowberry is 80 000 t, but only one tenth of these amounts is harvested. On average, 1 600 t of berries are picked annually. The average data on the seasonal development of the bilberry and cowberry obtained during long-term observations allow phenological forecasting of the optimal times for berry gathering. A method of short-term yield forecasting was elaborated. This will allow better exploitation of berry resources. Some proposals on the optimization of berry harvesting in Karelia are given.

Key words: Bilberry and cowberry resources, harvesting times, yield forecasting

*T.V. Belonogova, Forest Research Institute of the Karelian Branch of the USSR Academy of Sciences, ul. Pushkinskaja 11, SU-185610 Petrozavods, USSR*

In the food program of the USSR account is also taken of the use of wild food plants. Nearly 170 000 t of wild fruit and berries are picked annually in the country, but this does not cover the growing industrial and domestic demand. Intensive and long-term exploitation of food plants must be closely connected with their rational use. An inventory of the total resources and forecasting of the fruit yield are very important for efficient exploitation.

About 20 species of berry plants can be found in Karelia, but only a few are widely gathered — the bilberry (*Vaccinium myrtillus* L.), the cowberry (*V. vitis-idaea* L.) and the cranberry (*Oxycoccus palustris* Pers.), less frequently the cloudberry (*Rubus chamaemorus* L.) and the ashberry (*Sorbus aucuparia* L.), and in southern regions the raspberry (*Rubus idaeus* L.) and strawberry (*Fragaria vesca* L.). Of these species, the bilberry and cowberry represent considerable natural resources, being common in almost all types of taiga forests. According to the available data, the natural yield of the bilberry is 100 000 t and that of the cowberry 80 000 t, but their exploitation is one tenth of that amount. On average 1 600 t of berries are collected annually by some enterprises and organizations. The amounts vary greatly, however, from one year to another, depending on the yield. Wild berries are bought from the population through the network of storing places by workers from the storing organizations. Berries are usually gathered around the settlements and along the roads. Less accessible but more productive stands remain poorly exploited. Berry collection is economically advisable in Karelia. When the yield is good or average, the profit from one ton of cowberries may be 300–340 roubles, the corresponding value for the bilberry being 250–300 roubles.

## SEASONAL DEVELOPMENT OF THE BILBERRY AND COWBERRY

At present, a decrease of the berry resources has been observed in Karelia, which is connected with the intensification of silviculture and reduction of the areas with wild berries. So determination of the optimal collection period will promote more efficient use of the berry resources.

Study of the seasonal development of berry plants is important for forecasting the dates of mass ripening of the fruit and the value of their yield. Such observations have been made by the Forest Research Institute of the Karelian Branch of the USSR Academy of Sciences since 1970. Average data on the seasonal development of the bilberry and cowberry obtained during the long-term observations allow phenological forecasting, i.e. preliminary determination of the various stages. This enables us to determine the optimal dates for berry gathering, to prepare for the collection and to coordinate phenological forecasting with prediction of the yield, taking account of the occurrence of night frosts.

In Karelia the bilberry is most common in bilberry forest types (Tables 1 and 2). The studies showed that the greening of its buds begins almost immediately after the snow melts (14–

Table 1. *Phenological stages of bilberry development in Karelia in three different forest types (average long-term data).*

| Stages, substages | Bilberry pine stand Period | Bilberry long-stem moss birch wood Period | Bilberry spruce stand Period |
|---|---|---|---|
| Vegetative: | | | |
| Greening of leaf buds | (10–)**14**(–18).V | (13–)**17**(–21).V | (13–)**18**(–23).V |
| Unrolling of most leaves | (20–)**24**(–28).V | (22–)**27.V**(–1.VI) | (25–)**29.V**(–2.VI) |
| Budding: | | | |
| Swelling of flower buds | (13–)**18**(–23).V | (16–)**20**(–24).V | (16–)**21**(–26).V |
| Unrolling of flower buds | (16–)**20**(–24).V | (19–)**22**(–25).V | (21–)**24**(–27).V |
| Mass unrolling of flower buds | (21–)**25**(–29).V | (22–)**26**(–30).V | (25–)**29.V**(–2.VI) |
| Flowering: | | | |
| Opening of the first flower | (20–)**23**(–26).V | (21–)**25**(–29).V | (24–)**27**(–30).V |
| Mass flowering | (29.V–)**1**(–4).VI | (31.V–)**2**(–4).VI | (31.V–)**3**(–6).VI |
| End of flowering | (30.V–)**4**(–9).VI | (1–)**5**(–9).VI | (2–)**6**(–10).VI |
| Fruiting: | | | |
| Beginning of fruit setting | (1–)**5**(–9).VI | (2–)**6**(–10).VI | (4–)**7**(–10).VI |
| Mass fruit setting | (9–)**12**(–15).VI | (10–)**13**(–16).VI | (11–)**15**(–19).VI |
| Appearance of first mature fruit | (18–)**20**(–22).VII | (18–)**21**(–24).VII | (19–)**22**(–25).VII |
| Mass ripening | (25–)**28**(–31).VII | (28–)**30.VII**(–1.VIII) | (30.VII–)**1**(–3).VIII |

18 May). Leaf unrolling and budding are simultaneous. Flowering begins on 23–27 May and is completed on 4–6 June. Mass flowering begins 7–8 days after the opening of the first flower. The fruits set in early June and the first ripe berries appear after 20 July. Mass fruit ripening takes place in late July and early August.

The cowberry grows throughout Karelia, mainly in pine and spruce and rarely in birch forests, being most abundant in cowberry pine stands. Uneven distribution of the cowberry is observed on felling areas, where it can fruit successfully. Unlike the bilberry, the cowberry has both reproductive and vegetative buds. The vegetative buds open at the end of May, after

Table 2. *Phenological stages of cowberry development in cowberry pine stand and felling area of the same forest type (average long-term data).*

| Stages, substages | Cowberry pine stand Period | Felling area Period |
|---|---|---|
| Vegetative: | | |
| Bud swelling | (2–)**8**(–14).V | (2–)**8**(–14).V |
| Greening of leaf buds | (25–)**28**(–31).V | (14–)**21**(–28).V |
| Unrolling of most leaves | (10–)**13**(–16).VI | (24–)**29.V**(–3.VI) |
| Bud setting | (13–)**16**(–19).VII | (12–)**16**(–20).VII |
| Budding: | | |
| Swelling of flower buds | (4–)**9**(–14).V | (3–)**8**(–13).V |
| Unrolling of flower buds | (29.V–)**1**(–4).VI | (–18)**21**(–24).V |
| Mass unrolling of flower buds | (15–)**17**(–19).VI | (4–)**8**(–12).VI |
| Flowering: | | |
| Opening of the first flower | (12–)**15**(–18).VI | (10–)**13**(–16).VI |
| Mass flowering | (16–)**21**(–26).VI | (15–)**18**(–21).VI |
| End of flowering | (29.VI–)**2**(–5).VII | (27.VI–)**1**(–5).VII |
| Fruiting: | | |
| Beginning of fruit setting | (26–)**30.VI**(–4.VII) | (20–)**25**(–30).VI |
| Mass setting | (7–)**12**(–17).VII | (3–)**7**(–11).VII |
| Appearance of the first ripe fruit | (16–)**19**(–22).VIII | (5–)**10**(–15).VIII |
| Mass ripening | (30.VIII–)**3**(–7).IX | (20–)**27.VIII**(–3.IX) |

Table 3. *Periods of berry plant development (in days) of the cowberry and bilberry in different forest types.*

| | Beginning to end of flowering | End of flowering to appearance of first ripe fruit | Mass flowering to mass fruit ripening |
|---|---|---|---|
| Cowberry: | | | |
| Cowberry pine stand | 17 | 46 | 73 |
| Calamagrostidetum arundinaceae – lerchenfeldiosum felling area | 18 | 38 | 70 |
| Bilberry: | | | |
| Bilberry pine stand | 11 | 44 | 56 |
| Bilberry long-stem moss birch wood | 10 | 45 | 57 |
| Bilberry spruce stand | 9 | 45 | 57 |

which the growth of vegetative shoots becomes noticeable. It ends during the mass fruit ripening in late August. The generative buds are usually formed in July, on short shoots, and from the middle of August they can be distinguished from the vegetative ones. In the pine felling areas, the phenological stages begin some days earlier than in stands of the same forest type. Budding is prolonged and takes place in May–June.

The seasonal development was found to depend on the weather. When the meteorological conditions are unfavourable, the phenological stages begin somewhat later than the average dates given here, while favourable weather advances the start of a stage and shortens its duration.

The time at which a phenological stage begins can be calculated from the data in Table 3. The cowberry flowers for 17 days, and 63 days elapse between flowering and fruit ripening. In the bilberry the interval between mass fruiting and mass fruit ripening is 56–57 days.

The phenological forecast must be adjusted according to the data of the nearest meteorological stations.

The aim of our studies is to discover whether short-term forecasts of the bilberry and cowberry yields can be made from the number of flowers and ovaries. Such forecasting would be most advantageous in Karelia. Long-term observations have provided data allowing us to make such a forecast.

The material on flower and bud loss helps us to characterize the seasonal development of the bilberry and cowberry and to evaluate their potential yield (Table 4). An important factor for crop forecasting from flower number is the occurrence of night frosts during the flowering of berry plants. The conditions of the habitat are also very important. Development of bilberry and cowberry fruits from the flowers may vary with the conditions of the habitat from 0 to 50%. So the forecasts must be made individually for the berry stands.

Table 4. *Percentages of bilberry and cowberry flowers and ovaries yielding fruits in different types of forest (average long-term data).*

| Forest type (felling area) | Flowers | Ovaries |
|---|---|---|
| Bilberry: | | |
| Bilberry pine stand | 34 | 53 |
| Bilberry spruce stand | 18 | 45 |
| Bilberry birch wood | 24 | 50 |
| Calamagrostidietum arundinaceae – lerchenfeldiosum felling area | 24 | 50 |
| Cowberry: | | |
| Cowberry pine stand | 38 | 68 |
| Cowberry felling area | 18 | 46 |

Thus, to ensure the rational use of berry resources it is necessary to determine the most productive berry forests. Strict observance of the period of berry gathering and yield forecasting for the planning of the places and means of collection are very important. A phenological observation service should be organized in every silvicultural farm of the republic to make the forecasts. It is also necessary to determine the most efficient ways of using the berry resources according to the economic and forest-plant conditions of Karelia. This includes establishing a wide network of storing places, which must be near the areas of gathering. Special teams of workers should be formed and provided with transport. In future, it is proposed to carry out silvicultural measures directed at improving the productivity of the berry forests already in use.

Received 28.VII.1987

Acta Bot. Fennica 136:23–25, 1988

# Bilberry and cowberry yields and the factors controlling them in the forests of Karelia, USSR

A.A. KUCHKO

Kuchko, A.A. 1988: Bilberry and cowberry yields and the factors controlling them in the forests of Karelia, USSR. — In: Vänninen, I. & Raatikainen, M. (eds.), Proceedings of the Finnish-Soviet symposium on timber forest resources in Jyväskylä, Finland, 25–29 August 1986: 23–25. Acta Bot. Fennica 136. Helsinki. ISSN 0001-5369 ISBN 951-9469-30-3

The fruiting of the bilberry and cowberry were studied in 1974–1983 on 96 forest plots. The bilberry yield is more stable in pine stands (85–105 kg/ha), but fluctuates significantly in spruce (35–139 kg/ha) and birch (55–155 kg/ha) stands. Higher yields were observed in transition areas between dry and wetter forest types. The highest cowberry yields were observed in pine stands and cowberry spruce stands (100–120 kg/ha). The main factors decreasing the bilberry and cowberry yields are: late springs, night frosts, and cold rainy weather or dry soil during the ripening of the berries.

Key words: Bilberry, cowberry, fruiting

*A.A. Kuchko, Forest Research Institute of the Karelian Branch of the USSR Academy of Sciences, Pushkinskaya 11, SU-185610 Petrozavodsk, USSR*

## INTRODUCTION

The bilberry (*Vaccinium myrtillus* L.) and cowberry (*V. vitis-idaea* L.) are the most common berry plants in the Karelian forests. They can be found in all forest types but the cowberry has its ecological optimum in sparse pine stands on fresh sandy soils and the bilberry in fairly dense coniferous and deciduous stands, growing on sandy loams or loamy clays. Both species, especially the cowberry, can persist and give abundant fruit for some time after clear cutting. Their fruits and leaves are of great importance for alimentary and medicinal purposes and as food for wild animals and birds, and are widely gathered both privately and by state organizations.

The literature on the bilberry and cowberry in the USSR is very great, but in Karelia such investigations only began recently, in the late sixties. The first report on bilberry and cowberry productivity was made in 1969, in the Kivach Reserve by Yakovlev (1971). Wider, systematic study then began on berries and medicinal plants in Karelia, and by now there are a number of serious publications, dealing with various aspects of their growth, development, fruiting, berry yield, medicinal use, etc. A review has been made of the data obtained for the use of those collecting and storing the plants and berries (Belonogova et al. 1982).

## METHODS

The Forest Research Institute of the Karelian Branch of the USSR Academy of Sciences began to study bilberry and cowberry productivity in 1974. The study combines investigation of permanent sample plots with route inventories. The specially selected sample plots (0.3–0.4 ha each) are located in different forest types; the "random" routes cross different taxation portions. The sample plots are chosen with regard to the demands of forestry. A complete description of the ground cover and an assessment of the tree canopy are made on them.

The berry yield within the sample plots is assessed in 40 evenly spaced 1-m$^2$ quadrats. When the route investigations are carried out, the yield estimates are followed by a brief description of the main phytocenosic and topographic factors influencing the productivity. The berries are counted on temporary quadrats measuring 0.5 x 1 m. The quadrats number 30–40 to 100, depending on the size and character of the stands. The precision of the estimates is 15–20%, which is acceptable for both industrial and scientific purposes (Pozdnyakov 1973).

In every sample plot 100 berries were selected and weighed in 3–5 replicates for the yield determination, and the yield per hectare was calculated.

## RESULTS AND DISCUSSION

I here present the results of a 10-year (1974–1983) period of investigations. My material is supplemented with data of my colleagues T.V. Belonogova, T.G. Voronova, N.L. Zaitseva and V.S. Sakovets.

The fruit yield of the bilberry, and other forest berries as well, is known to vary greatly from year to year. In four of the

Table 1. *Bilberry productivity in the forests of Southern Karelia (age class III and more, density 0.5–0.8).*

| Forest type | Number of plots | Aver. yield (kg/ha) for 6 years | Variation in yield (kg/ha) Between years | Between plots |
|---|---|---|---|---|
| Cowberry–bilberry pine stand | 8 | 90 | 30–170 | 10–265 |
| Bilberry pine stand | 10 | 85 | 45–140 | 10–185 |
| Moist bilberry pine stand | 4 | 105 | 35–140 | 30–165 |
| Cowberry–bilberry spruce stand | 4 | 35 | 10–50 | 5–60 |
| Bilberry spruce stand | 10 | 60 | 25–80 | 10–190 |
| Moist bilberry spruce stand | 5 | 130 | 30–230 | 30–250 |
| Bilberry birch stand | 10 | 55 | 10–90 | 5–120 |
| Moist bilberry birch stand | 6 | 155 | 30–240 | 10–350 |

study years the yield was poor, in three average and in three good. The yield depends on the weather in both the current and the previous year. A good yield requires favourable conditions for the formation of the generative buds (heat, moisture) and for their flowering and pollination, and ripening of the fruits. The decisive factor is the presence or absence of frosts during the flowering.

According to the long-term averages, bilberry flowering begins on 27 May (Romanovskaya 1969) and continues for 9–18 days. The probability of frost is fairly high in this period; in one of four years frost occurs till the middle of June, the 25th of June being the latest date (Spravochnik po klimatu SSSR 1965 (Manual of the climate of the USSR). All the study years with poor berry yields had late-spring frosts. Even without frosts, cold weather during the bilberry flowering decreases the activity of insect pollinators and thus reduces the berry yield.

Table 1 presents data on the bilberry productivity during the years with good and average crops in the most favourable forest types. The average productivity during 6 years was not higher than 150 kg per hectare. The bilberry yield is more stable in pine stands (85–105 kg/ha) and fluctuates widely in spruce stands (35–130 kg/ha). An increase in yield with soil moisture is observed in all forest formations within the bilberry forest types, but in too moist forest types (*Ledum palustre, Sphagnum*) the bilberry yield falls sharply, to 35 kg/ha.

The yields fluctuate very significantly within the forest types from one year to another and from one plot to another. In the last case the difference can be 20–30-fold. These fluctuations indicate the great influence of phytocenotic factors on the bilberry yield, including the canopy density, the abundance of seedlings and undergrowth, the projective berry bush cover, and the stand composition. The closure of the crown canopy, which influences the illumination, is closely correlated with the stand density. Density is determined more easily and exactly than the crown closure and is the index most often used.

The investigations show that conditions are optimal for the berry yield when the density is 0.5–0.8. In pure pine and birch stands it should preferably be 0.7–0.8 and in mixed and spruce stands 0.5–0.6. When the stand densities (and crown closure) are equal, the bilberry yields are higher in areas with fewer seedlings and less undergrowth. According to Lukin's data (1976), in the presence of dense seedlings and undergrowth, berry bushes receive less illumination under a canopy of low density (up to 0.5) than in very dense tree stands (0.8– 0.9). The seedlings and undergrowth also decrease access of heat to the field layer.

Canopy thinning connected with silvicultural or other measures promotes some increase in berry yield. In the areas with intermediate cuttings of high intensity the bilberry yield reached 250–300 kg/ha (Zaitseva & Belonogova 1981, Sakovets 1985). The positive effects of cutting on the productivity last for 7–10 years.

The highest bilberry yields are observed along the edges of clear-cut areas, particularly during the first 2–3 years after felling. Under these conditions they can reach 500 kg/ha and more (Sakovets 1985). The "highly productive bilberry strip" along the clear-cut boundary is, however, not more than 40–50 m wide.

The productivity of the cowberry also varies greatly, depending on the conditions of the habitat and weather. Flowering begins 10–15 days later than in the bilberry, so the night frosts are less dangerous to it, but in the years when the night frosts are late, the flower loss can reach 90–100%. Prolonged rains accompanied by low temperatures are very dangerous to the cowberry, and when the weather is warm and dry, 10–20% of the flowers can be damaged by caterpillars. During the berry ripening, great numbers of flowers, sometimes up to 50%, are shed because of soil dryness, or are eaten by animals (Belonogova & Rumyantseva 1985). Of the 10 study years, three had poor yields, five average yields and only two good yields.

Some data for seven years on stands of age class III and higher, with a density range of 0.4–0.9 are given in Table 2. As in the bilberry, productivity depends greatly on the year, and the forest type and area. The highest yields were observed in the cowberry forest types, and the average yields in pine and spruce stands differ slightly, though the difference may be two-fold in some plots. In pure cowberry pine stands, differences in stand density slightly influence the berry yield, but the decrease of the proportion of pine in the tree stand from 90 to 60% reduces the crop by a factor of 1.5–3. In the cowberry spruce stands, yields sufficiently large for exploitation are formed when the density is not higher than 0.6. In the bilberry forest types, the cowberry yield is one third to a quarter of that in the cowberry forest types.

On the whole, cowberry productivity under the forest canopy in Karelia is rather low, and industrial exploitation of berries is inadvisable. More promising in this respect were gaps in the forest and old (3–8 years) felling sites, where the average yields were 210 and 320 kg/ha, respectively (Zaitseva & Belo-

Table 2. *Cowberry productivity in the forests of Southern Karelia.*

| Forest type | Number of plots | Aver. yield (kg/ha) for 6 years | Variation in yield (kg/ha) | |
|---|---|---|---|---|
| | | | Between years | Between plots |
| Cowberry pine stand | 11 | 120 | 30–230 | 20–520 |
| Bilberry pine stand | 6 | 35 | 0–50 | 0–65 |
| Cowberry spruce stand | 6 | 100 | 10–170 | 10–220 |
| Bilberry spruce stand | 7 | 30 | 0–40 | 0–70 |
| Bilberry–cowberry birch wood | 6 | 45 | 10–80 | 0–120 |
| Bilberry birch wood | 8 | 10 | 5–20 | 0–30 |

nogova 1981). The cowberry is a luciphilous plant and grows well throughout a felling site. The period of its intensive development and fruiting is limited by the time of forest cutting and reforestation measures. Under optimal conditions, the cowberry yields on the felling sites may reach 1 500–2 000 kg/ha.

The berry size is of great economical importance and varies among forest types and plots (Table 3). Bilberries vary more in size: the variation in their average weight with the type of habitat is 26%, against 20% in the cowberry.

The largest bilberries are found in moist pine and birch stands (weight of 100 berries up to 38–40 g), the largest cowberries in gaps in the forest and in felling sites (up to 30 g).

The productivity of both the bilberry and the cowberry is thus closely connected with the meteorological conditions and the forest type, and within the type with the crown canopy closure. The yield fluctuates significantly between plots, even in the same forest type, and the difference may reach one–two orders. The largest bilberry crops are obtained from moist bilberry pine, spruce and birch stands and the edges of felling sites. The cowberry is most productive in sparse to fairly dense cowberry pine stands and clearings in cowberry and bilberry forests 3–8 years after felling.

Table 3. *Fruit size in bilberry and cowberry (mean weight and S.E. of 100 berries.*

| Habitat type | Weight (g) of 100 berries ± S.E. Bilberry | Cowberry |
|---|---|---|
| Cowberry pine stand | 32.5 ± 4.7 | 23.3 ± 4.2 |
| Bilberry pine stand | 33.9 ± 4.1 | 20.2 ± 5.0 |
| Moist bilberry pine stand | 36.3 ± 5.6 | – |
| Cowberry spruce stand | 25.3 ± 7.2 | 22.2 ± 7.2 |
| Bilberry spruce stand | 28.2 ± 5.3 | 22.8 ± 3.3 |
| Bilberry birch wood | 30.1 ± 7.5 | 24.7 ± 8.1 |
| Clear cutting | 27.0 ± 8.6 | 25.5 ± 7.7 |

## REFERENCES

Belonogova, T.V., Voronova, T.G., Zaitseva, N.L., Kuchko, A.A. & Sakovets, V.I. 1982: Uchët i ispol'zovanie resursov polesnykh rastenii lesov yuzhnoi Karelii. — 38 pp. Petrozavodsk.

Belonogova, T.V. & Rumyantseva, L.G. 1985: Formirovanie urozhaya cherniki i brusniki v lesakh Yuzhoi Karelii. — In: Sistema lesokhozyaistvennykh meropriyatii v sosnovykh lesakh Karelii: 128–137. Petrozavodsk.

Lukin, I.N. 1976: Rol' pologa drevostoya, podrosta i podleska v postuplenii sveta i tepla k yarusu yagodnykh kustarnichikov. — In: Voprosy lesovosstanovlenya na Evropeiskom Severe: 127–131. Arkhangelsk.

Pozdnyakov, L.K. 1973: Lesnoe resursovedenie. — 120 pp. Novosibirsk.

Romanovskaya, M.M. 1969: Kalendar' prirody zapovednika "Kivach". — In: Trudy zapovednika "Kivach": 82–86. Petrozavodsk.

Sakovets, V.I. 1985: Vliyanie rubok glavnogo pol'zovaniya na produktivnost' dikorastushchikh yagodnikov v lesakh Karelii. — In: Sistema lesokhozyaistvennykh meropriyatii v sosnovykh lesakh Karelii: 137–141. Petrozavodsk.

Spravochnik po klimatu SSSR. III. 1965. — 240 pp. Leningrad.

Yakovlev, F.S. 1971: Plodonoshenie cherniki i brusniki v sosnyakakh yuzhnoi Karelii. — In: Lesnye rastitel'nye resursy yuzhnoi Karelii: 136–139. Petrozavodsk.

Zaitseva, N.L. & Belonogova, T.V. 1981: Urozhainost' dikorastushchikh yagodnikov na vyrubkakh yuzhoi Karelii. — In: Resursy nedrevesnoi produktsii lesov Karelii: 41–49: Petrozavodsk.

Received 20.VII.1987

Acta Bot. Fennica 136:27–31, 1988

# The structure of South Karelian (USSR) cowberry coenopopulations

TAIMI PAAL

Paal, T. 1988: The structure of South Karelian (USSR) cowberry coenopopulations. — In: Vänninen, I. & Raatikainen, M. (eds.), Proceedings of the Finnish-Soviet symposium on timber forest resources in Jyväskylä, Finland, 25–29 August 1986: 27–31. Acta Bot. Fennica 136. Helsinki. ISSN 0001-5369 ISBN 951-9469-30-3

Data on nine phenological phases collected over seven years were used to construct a temperature-phenological nomogram. This nomogram makes it possible to analyse the dependence of interphasal periods on the air temperature of the current year, and to predict differences in phenological development between coenopopulations. Fruiting stability revealed that certain parts of coenopopulations have a tendency to be more fruitful than others. The phytogenic fields of trees exert only a weak influence on the cowberry yield.

Key words: Cowberry yield, phenology, phytogenic fields, *Vaccinium vitis-idaea*

*Taimi Paal, Laboratory of Nature Conservation, Estonian Research Institute of Forestry and Nature Conservation, 2 Rõõmu Road, SU-202400 Tartu, Estonian SSR, USSR*

To simplify the study of biological structures, it has been customary to adopt either a spatial or a functional approach (e.g. Beklemyshev 1960, Mazing 1969). According to Vasil'evich (1972), structural studies of the first type strive to define how the object is organized, studies of the second type are interested in how it works, or the process taking place within it. In this paper we shall examine some functional aspects of the cowberry yield. The field research was performed in 1976–1982 in the Kivach Nature Reserve in South Karelia.

## TEMPERATURE AND PHENOLOGY

Three sample plots were chosen whose measurement and topography corresponded to the habitats of certain coenopopulations:

- No. 4. *Vaccinium* type pinewood, age of tree stand 130 years, its density 0.6, and the projective cowberry cover 35%.
- No. 9. *Ledum–Sphagnum*-type pinewood (edge of raised pine bog), age of tree stand 140 years, density 0.8, cowberry cover 15%.
- No. 10. *Polytrichum*-type mixed pine-spruce forest, 180 years, density 1.0, cowberry cover 32%.

Within each plot, 60 cowberry bushes were chosen at random and marked. The dates of the phenological phases were recorded by the method of Batmanov (1966). Observations were made over 1–4 days, depending on the weather and the rate of cowberry development. The average dates of some phenophases are given in Table 1.

Table 1. *The phenology of the cowberry in 1976–1982 in the sample plots.*

| Sample plot | The dates of the phenophases ± error in days | | | | | |
|---|---|---|---|---|---|---|
| | Generative buds | | Mass flowering | | Ripening of berries | |
| | Swelling | Opening | Beginning | End | Beginning | End |
| 4 | 19.V. ± 2.4 | 27.V. ± 1.9 | 20.VI. ± 2.4 | 02.VII. ± 4.3 | 16.VIII. ±2.3 | 28.VIII ± 3.9 |
| 9 | 19.V. ± 2.6 | 26.V. ± 1.7 | 19.VI. ± 2.9 | 29.VI. ± 3.3 | 10.VIII. ± 2.3 | 23.VIII. ± 3.3 |
| 10 | 25.V. ± 3.0 | 31.V. ± 2.4 | 27.VI. ± 4.1 | 07.VII. ± 3.3 | 18.VIII. ± 5.8 | 05.IX. |

A thermal-phenological nomogram was constructed (Podol'skii 1974) for forecasting the dates of the flowering and ripening of the cowberry, and elucidating the relationship of the phenology with the thermal climate of a given region. The initial date of the growing season was taken as May 1. The relative error of the mean air temperatures of the 10-day periods from the 1st third of June to the 2nd third of September (in 1974–1984) was less than 7%. Using the data of every sample plot separately, empirical curves were constructed describing the relationship between the air temperature and the length of the interphasal periods in cowberry development (Fig. 1).

Graphic interpolation of the network of thermal data within the studied region makes it possible to find the mean temperature of any period commencing on the initial date; interpolation of different lines in the network makes it possible to choose any initial date. Inverse checking of the exactness of the beginnings of different phenophases of the cowberry on the thermal phenological nomogram demonstrated that the predictive error averaged 3–8 days. This may be regarded as a satisfactory result.

The extent of the base of the network, i.e. the abscissa, gives an idea of the amplitude of variation of the 10-day temperature means during a certain growth period in the given climatic region. The interval on the abscissa between the ends of the phenocurves shows the limits of the mean air temperature in a certain interphasal period in the given environmental conditions. It is apparent that the interphasal periods connected with the beginning of the ripening of the berries take place in a relatively limited interval of the mean temperatures.

The steepness of the phenocurves expresses the influence of the air temperature upon the rate of development of the plant: the steeper the phenocurve, the more sensitive is the plant in the given interphasal period to changes of temperature. The figure demonstrates that the length of the period from swelling of the generative buds to opening of the buds does not shorten much with rising temperature. The phenocurves are more or less parallel with the axis of the abscissa. It is the period between swelling of the buds and the beginning of mass flowering that depends most on the temperature: a rise by 3°C — from 10 to 13°C — shortens the period by up to 15–25 days in different biotopes (Fig. 1). It should be noted, however, that care must be exercised in interpreting the steepness of the phenocurves; the periods during which the plants depend most strongly on this or that environmental factor may not always correspond to the various phenophases.

The differences between the phenocurves describing the same interphase period in different habitats are also rather interesting. The curves are located quite close to one another and are

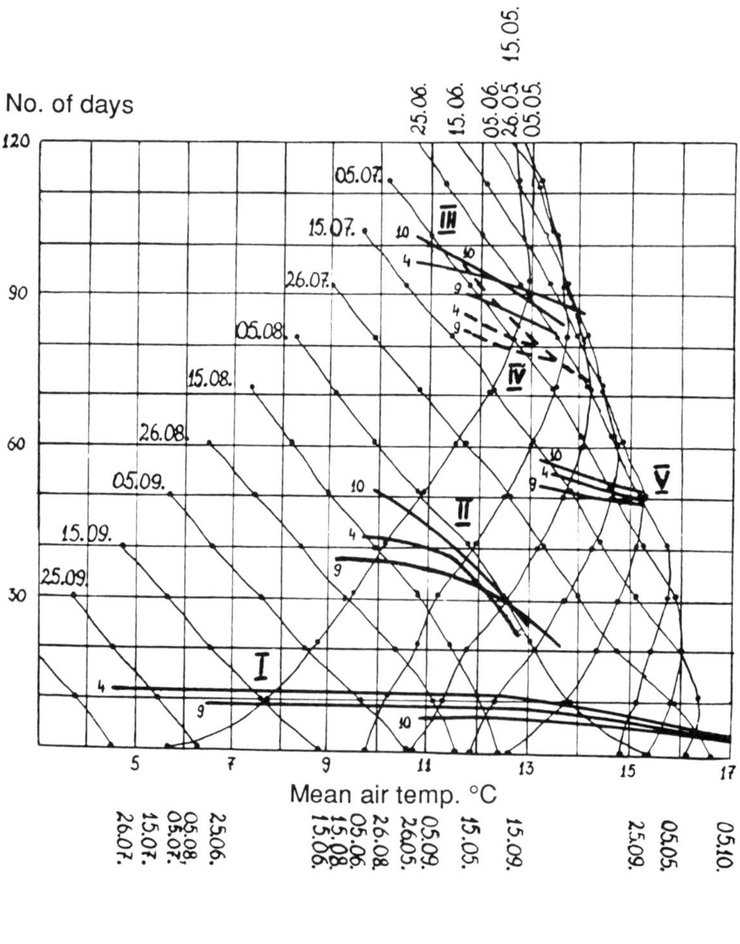


*Fig. 1.* The thermal–phenological nomogram of cowberry development. I–V = interphasal phenological periods I) swelling of the generative buds – opening of the generative buds; II) swelling of the generative buds – beginning of mass flowering; III) swelling of the generative buds – beginning of the ripening of the berries; IV) opening of the generative buds – beginning of ripening of the berries; V) beginning of mass flowering – beginning of ripening of the berries. The corresponding curves in are marked with roman numerals: the Arabic numerals denote the sample plots.; 4, 9, 10 = the phenocurves of the separate coenopopulations; 05.5.–05.10. = dates.

of similar character, but some differences are still observable. It can be concluded from the nomogram that a cowberry coenopopulation in a *Polytrichum*-type forest (sample plot No. 10) is more sensitive to the temperature than the populations in other habitats during the periods between swelling of the buds and the beginning of ripening of the berries. The same is also true of the period between opening of the buds and the beginning of ripening of the berries. In the population in *Ledum–Sphagnum*-type pinewood (No. 9) the period from swelling of the buds to the beginning of ripening of the berries is shorter than in the other coenopopulations. These differences in the phenocurves reflect the special features of the habitat and the deviation of its temperature regime from the mean data of the meteorological station.

## THE STABILITY OF FRUITING IN DIFFERENT PARTS OF THE COENOPOPULATIONS

The studies were performed in 1976–1978 on the same sample plots and on one additional plot (No. 5) located on a west-facing slope of 5–6°: *Vaccinium*-type pine stand, age 130 years, density 0.6 and cowberry cover 46%.

The average cowberry yield was estimated on randomly chosen 1 $m^2$ subplots (Table 2). The estimates can be viewed as parameters of the yield from different clones or their parts (Paal et al. 1983).

In the study period the average yield varied especially greatly in dry cowberry pinewoods (sample plots Nos. 4 and 5), being more stable in damp habitats; the distribution of the berries was more even on the productive sample plots, because the number of subplots completely lacking berries was smaller.

For the quantitative analysis of the stability of fruiting in the different parts of the coenopopulations, 40 estimates were chosen at random for every sample plot from the whole set of data for 1976. The subplots without cowberries were not taken into account. The estimates were then ranked separately for each sample plot and the corresponding estimates for the following years were ranked according to the order of the first year of study. Each of these rankings was divided into 8 groups, every group comprising the estimates of 5 subplots for 3 years. The data of each year was processed separately using a two-factor hierarchical analysis of variance according to the following model:

$$X = \mu + P + S(P) + \varepsilon,$$

where X is the yield of berries, $\mu$ is the average yield from all the plots, P reflects the differences between sample plots (the effect of the "plot" factor), S(P) reflects the differences in yield estimates between groups of subplots within a certain plot (effect of the hierarchical factor) and $\varepsilon$ is the random error. According to the results of all the three analyses, the factors considered, i.e. the plots and the groups of subplots within a certain plot, appeared to have a significant effect on the estimate of the yield (Table 3).

The regression coefficients for the levels of the hierarchical factor in the analyses of variance can be regarded as estimates of the relative fruitfulness of the groups of subplots. Negative coefficients denote yields smaller than the average of the plot, positive values larger yields than average. The stability of the difference in fruitfulness between the subplot groups can be estimated by calculating the correlation between the regression coefficients for the same levels of the hierarchical factor in successive years (Table 4).

The correlation coefficients show that on plots 5 and 9 the difference in fruiting between parts of the populations is fairly constant; i.e. the high-yielding (or low-yielding) subplots were the same in all three years. The difference appeared to be less constant in plot 4, and varied in a random manner in plot 10, due to the small quantity of berries. Although not all the correlation coefficients appeared significant, this is not necessarily evidence that interdependence does not exist. It may be caused by the short row of compared variables. The multiple correlation of the regression coefficients of the hierarchical factor of all three variance analyses for the different sample plots was: 0.726 (No. 4), 0.823 (No. 5), 0.976 (No. 9) and 0.235 (No. 10) (Table 4).

From this we may conclude that the fruiting of cowberry populations is not distributed randomly in different years: there is a tendency for certain parts of the populations to be markedly more fruitful than the others (in the absence of pests and unfavourable weather). This should be born in mind in choosing plants for cultivation and selective breeding.

Table 2. *The average cowberry yield on 1 $m^2$ subplots in 1976–1978.* S.E. = error of the mean, S.D. = standard deviation, C.V. = coefficient of variation.

| Sample plot | Year | No. of subplots | Characteristics of yield | | | |
|---|---|---|---|---|---|---|
| | | | Mean, g | S.E. | S.D. | C.V. |
| 4 | 1976 | 80 | 1.7 | 0.24 | 2.14 | 1.26 |
| | 1977 | 80 | 12.4 | 1.09 | 9.75 | 0.79 |
| | 1978 | 100 | 10.2 | 0.75 | 7.54 | 0.74 |
| 5 | 1976 | 80 | 4.6 | 0.63 | 5.63 | 1.22 |
| | 1977 | 84 | 23.6 | 2.02 | 18.52 | 0.78 |
| | 1978 | 80 | 19.3 | 1.72 | 15.35 | 0.80 |
| 9 | 1976 | 80 | 3.1 | 0.65 | 5.78 | 1.86 |
| | 1977 | 80 | 8.3 | 1.62 | 14.50 | 1.75 |
| | 1978 | 80 | 5.0 | 1.01 | 8.99 | 1.80 |
| 10 | 1976 | 40 | 0.3 | 0.19 | 1.19 | 3.97 |
| | 1977 | 40 | 0.6 | 0.21 | 1.33 | 2.22 |
| | 1978 | 40 | 0.4 | 0.11 | 0.67 | 1.68 |

Table 3. *The coefficients of hierarchical two-factor analysis of variance applied to the estimates of the cowberry yields in different years.*

| | Level of factor | Coefficients 1976 | 1977 | 1978 |
|---|---|---|---|---|
| Average yield from all the plots | | 2.16 | 10.73 | 8.69 |
| Sample plot | | | | |
| | 1 | –0.90 | –0.87 | 1.11 |
| | 2 | 1.23 | 14.21 | 11.35 |
| | 3 | 1.51 | –3.24 | –4.14 |
| | 4 | –1.84 | –10.10 | –8.32 |
| Groups of subplots in the sample plot | | | | |
| | 1(1) | –1.27 | –9.06 | –7.78 |
| | 2(1) | –1.27 | –7.46 | –6.08 |
| | 3(1) | –1.27 | –4.26 | 0.58 |
| | 4(1) | –1.27 | 1.14 | 6.20 |
| | 5(1) | –0.53 | 7.64 | –0.98 |
| | 6(1) | 0.20 | 12.54 | 5.68 |
| | 7(1) | 1.74 | 0.52 | 2.55 |
| | 8(1) | 3.66 | –1.06 | –0.16 |
| | 1(2) | –3.40 | –24.74 | –18.74 |
| | 2(2) | –3.40 | –16.34 | –5.96 |
| | 3(2) | –3.40 | 6.46 | 3.80 |
| | 4(2) | –1.96 | 10.96 | 6.80 |
| | 5(2) | –1.26 | –0.24 | 9.30 |
| | 6(2) | 0.83 | 5.76 | –7.00 |
| | 7(2) | 2.83 | 1.26 | –0.12 |
| | 8(2) | 9.75 | 16.86 | 11.90 |
| | 1(3) | –3.67 | –7.49 | –4.55 |
| | 2(3) | –3.67 | –7.49 | –4.55 |
| | 3(3) | –3.67 | –7.49 | –3.99 |
| | 4(3) | –3.67 | –2.79 | –0.49 |
| | 5(3) | –2.89 | –1.59 | –2.93 |
| | 6(3) | –0.83 | –3.79 | –2.85 |
| | 7(3) | 3.05 | 16.61 | 2.55 |
| | 8(3) | 15.37 | 14.01 | 16.83 |
| | 1(4) | –0.33 | –0.63 | –0.37 |
| | 2(4) | –0.33 | –0.63 | –0.37 |
| | 3(4) | –0.33 | –0.63 | –0.37 |
| | 4(4) | –0.33 | –0.63 | 0.29 |
| | 5(4) | –0.33 | –0.63 | 1.43 |
| | 6(4) | –0.33 | –0.13 | –0.27 |
| | 7(4) | –0.33 | 1.08 | –0.17 |
| | 8(4) | 2.28 | 2.18 | –0.23 |

## DEPENDENCE ON THE TREE CANOPY OF COWBERRY YIELDS

One of the chief elements in the characteristics of the horisontal structure of forest populations is the stem density of the tree stand, which determines the compactness of the layers of the tree canopy (Uranov 1965). This, in turn, is one of the chief factors determining the arrangement and character of the undergrowth.

The studies were undertaken in 1977–1978 in three *Vaccinium*-type pine stands (plots 4, 7, 8), differing in density, but more or less similar in floristic, phytocoenotic and soil characteristics (Paal et al. 1981). The density of the tree stands in the different plots were: No. 4: 0.6, No. 7: 0.7, and No. 8: 0.8. To estimate the cowberry yield, subplots of 0.5 x 0.5 m were placed along transects between the trees. Random pairs of trees were found by the nearest neighbour method (Cottam & Curtis 1949) and by the wandering quarter method (Catana 1963).

First, one-factor analysis of variance was used, the distance of subplots from the nearest tree trunk being taken as the determining factor. The data for 1977 were pooled, regardless of

Table 4. *The regressions of the phenological and morphological characters on latitude (LAT) and longitude (LONG) (in the two morphological characters also on AGE of the stem).* HEIGHT = height of the stems, GR86 = length of the current shoots in 1986.

| Phenological event | Significance of regression on LAT | LONG | Regression equation | Percentage of variation explained by regression |
|---|---|---|---|---|
| ONSET–LBI | *** | – | 3.018 LAT + 145.222 | 4.0 |
| ONSET–LBII | *** | – | 2.918 LAT – 103.084 | 3.8 |
| GB–B | – | ** | 5.384 LONG + 212.878 | 2.9 |
| LBII–CG | – | * | 1.795 LONG + 169.097 | 0.8 |
| B–BAC | – | * | –0.481 LONG + 32.307 | 1.6 |

| Morphological character | LAT | LONG | AGE | Regression equation | Percentage of variation explained by regression |
|---|---|---|---|---|---|
| HEIGHT | *** | * | *** | –1.095 LAT + 1.783 AGE + 0.474 LONG + 67.309 | 32.8 |
| GR86 | *** | – | ** | –1.801 LAT – 2.296 AGE + 164.604 | 4.6 |

their plot or method of collection. Distance was measured on an 11-degree scale, the greatest distance from the nearest tree being 5.5 m.

A probable influence of the factor on the cowberry yield was observed only for the two first degrees on the scale (0.0–0.5 m and 0.5–1.0 m from the trunk), where there were fewer berries than on the more distant subplots. However, distance served to explain only 4.7% of the total variation or, in other words, the influence of the tree canopy on the cowberry yield is relatively weak at a distance of 1.0 m and is totally lacking at a greater distance.

The above result was confirmed when covariant three-factor analysis of variance was applied to the data for both years, using methods and sample plot as additional factors. Only 1.5% of the total variation was related to the distance factor. The variation was best explained by interdependence of the yields in 1977 and 1978, which accounted for 10%.

The weak influence of the tree canopy on the cowberry yield indicates possible difficulties in elaborating methods of determining wild berry yields on the basis of the species composition of the tree stand. This question needs further investigation, however, as during the study years the cowberry yields in the Nature Reserve were being markedly diminished by pests (Paal et al. 1982).

## REFERENCES


Batmanov, V.A. 1966: Integral'nii i ekonometricheskii metody fenologicheskikh issledovanii. — Doklady fenologicheskogo sektora VGO 2(18): 49–65.

Beklemyshev, V.N. 1960: Prostranstvennaya i funktsional'naya struktura populjatsii. — Byull. MOIP, nov. ser., otd. biol. 65(2): 41–50.

Catana, A.J. 1963: The wandering quarter method of estimating population density. — Ecology 44:348–360.

Cottam, G. & Curtis, J.T. 1949: A method for making rapid surveys of woodlands by mean of pairs of randomly selected trees. — Ecology 30:101–104.

Mazing, V.V. 1969: Teoreticheskie i metodicheskie problemy izucheniya struktury rastitel'nosti. — Dokl. po opubl. rabotam, predstavlennyi k zashchite na soiskanie uch. stepeni doktora biol. nauk. 96 pp. Tartu.

Paal, T.V., Kutenkova, N.N. & Martin, M.O. 1982: Nasekomye-vrediteli Vaccinium vitis-idaea L. v Karelii. — Rastitel'nye resursy 18:411–417.

Paal, T.V., Paal, Ya. L. & Harin, V.N. 1981: Vliyanie fitogennykh polei derev'ev na urozhainost' brusniki. — Rastitel'nye resursy 17:68–74.

Paal, T.V., Paal, Ya. L. & Harin, V.N. 1983: Ob otsenke postoyanstva plodonosheniya Vaccinium vitis-idaea L. — Rastitel'nye resursy 19:118–124.

Podol'skii, A.S. 1974: Fenologicheskii prognoz. — 287 pp. Moscow.

Uranov, A.A. 1965: Fitogennoe pole. — In: Problemy sovremennoi botaniki: 183–208. Moscow–Leningrad.

Vasil'evich, V.I. 1972: Kolichestvennye metody izucheniya struktura rastitel'nosti. — Itogi nauki i tekhniki. Ser. bot. I: 7–83. Moscow.


Received 20.VII.1987

Acta Bot. Fennica 136:33–36, 1988

# Biology and berry production of the cowberry in Estonian SSR

R. MÄNNI

Männi, R. 1988: Biology and berry production of the cowberry in Estonian SSR. — In: Vänninen, I. & Raatikainen, M. (eds.), Proceedings of the Finnish-Soviet symposium on timber forest resources in Jyväskylä, Finland, 25–29 August 1986: 33–36. Acta Bot. Fennica 136. Helsinki. ISSN 0001-5369 ISBN 951-9469-30-3

The berry production of the cowberry depends on the cover of the berry plants, the density of co-dominants, and the density and height of the tree stand, the maximum correlation coefficient being 0.81. In different phytocoenoses, the same coenopopulations and clones within one coenopopulation always give better crops. From each blossom bud there develop on average 5.63 blossoms, from each blossom 0.49 ovaries, and from each ovary 0.58 berries. Self-pollination takes place side by side with insect pollination. The germination of seeds in the laboratory was 55%. The different phenophases of the cowberry can be predicted by temperature sums over 0°C.

Keywords: Berry production, *Vaccinium vitis-idaea*

*R. Männi, Estonian Forest Management Enterprise, Sutiste tee 36–70, Su-200034 Tallinn, Estonian SSR, USSR*

## MATERIAL AND METHODS

The investigations were carried out in 1978–1984 in 24 subplots in pine forests of the Cladonia–Calluna type in North-West Estonia and on five subplots in Vaccinium-type pine forests in South-East Estonia, altogether on 430 and 100 permanent quadrats (1 m$^2$), respectively. The mean data for the plots in the Cladonia–Calluna-type and, in brackets, the Vaccinium-type pine forests were the following: tree stand — pine 10, height 12 (22) m, diameter 20 (24) cm, age class 5.5 (4.8), site class 5.2 (2.4), stand density 0.40 (0.61), crown coverage 0.41 (0.46), number of trees per hectare 480 (460); type of soil — podzol, depth of organic soil horizon 4.9 (4.4) cm, soil fertility 1.7 (3.3), moisture content 2.1 (2.8); relief slightly undulating, microrelief level, cover of tree canopy 42 (46) %, cover of cowberries 37 (37) %, mean height of cowberry layer 9 (9) cm, lichen-moss cover 67 (94) %, density of co-dominants (*Empetrum nigrum* L., *Calluna vulgaris* L., *Vaccinium myrtillus* L.) 3.1 (4.7) points, cover 40 (7) %. The points express the percentage of co-dominants in direct competition with cowberry, according to a 6-point scale: 5 = co-dominants missing or occur as scattered individuals; 4 = 5–14% of cowberries; 3 = 15–34%; 2 = 35–54%; 1 = 55–74%; 0 = over 74%.

A similar scale was used for ocular estimation of productivity: 0 = blossoms, ovaries, berries are imperceptible; 1 = 1–14 % of shoots have blossoms, ovaries or berries; 2 = 15–34% of shoots; 3 = 35–54%; 4 = 55–74%; 5 = more than 74%.

## RESULTS

### Factors influencing the productivity of the cowberry

A survey of productivity in different years and sample plots is given in Table 1. The sample plots were formed by grouping subplots which lie close to each other, and have a similar crown coverage and phytocoenosis. Productivity varies not only with the year but also with the region: in the same year the crop of one region may be several times as great as that of another. The better the year or the more fertile a sample plot is, the less the crop varies and the smaller is the variation between an equal number of observations. Comparison of the number of berries and the points for productivity shows that the potential productivity is significantly higher. In 1978 a record sample plot gave 543 berries per m$^2$, in 1981 the same plot gave 1 047 berries per m$^2$.

Stepwise regression analysis was applied to the data about the Cladonia–Calluna type pine forests in the different years, the limit for inclusion being a partial correlation value of 0.15. The four following factors most often remain in the equation: cover of berry-plants, density of co-dominants, crown coverage and height of the tree stand. The correlation coefficient between the number of berries and the four main factors was 0.47, while the correlation coefficient between the number of berries and all the factors investigated was 0.51. So it is possible to describe

*Table 1. Berry production in the different years and sample plots.*

| Pine forest type | Year or sample plot | Mean number of berries/m² | Standard deviation (S.D.) | Coefficient of variation (S.D./mean, %) | Standard error of the mean (S.E./mean, %) | Productivity rating (ocular) 0–5 | Weight of one berry, g |
|---|---|---|---|---|---|---|---|
| | Year | | | | | | |
| Cladonia–Calluna | 1978 | 185.3 | 169.0 | 91 | 4.6 | 3.1 | 0.20 |
| | 1979 | 64.0 | 85.5 | 133 | 6.7 | 1.7 | 0.21 |
| | 1980 | 33.9 | 54.6 | 161 | 7.8 | 1.6 | 0.18 |
| | 1981 | 112.7 | 160.7 | 142 | 6.9 | 2.2 | 0.22 |
| | 1982 | 85.2 | 121.3 | 142 | 6.9 | 1.9 | 0.16 |
| | 1983 | 25.2 | 34.6 | 136 | 6.6 | 1.2 | 0.19 |
| | 1984 | 119.1 | 29.7 | 108 | 5.3 | 2.5 | 0.20 |
| | 1978–1984 | 88.6 | 128.1 | 144 | 2.7 | 2.1 | 0.19 |
| Vaccinium | 1981 | 11.9 | 18.5 | 155 | 15.6 | 1.2 | |
| | 1982 | 97.4 | 102.2 | 105 | 10.5 | 2.0 | |
| | 1983 | 133.2 | 136.0 | 102 | 10.2 | 1.8 | |
| | 1984 | 71.2 | 95.9 | 134 | 13.5 | 1.6 | |
| | 1981–1984 | 78.4 | 107.3 | 136 | 8.7 | 1.7 | |
| | Sample plots | | | | | | |
| Cladonia–Calluna | 1 (210)* | 14.3 | 30.8 | 215 | 15.0 | 0.8 | 0.16 |
| | 2 (219) | 99.3 | 114.4 | 115 | 8.0 | 2.5 | 0.20 |
| | 3 (420) | 86.7 | 115.2 | 132 | 6.5 | 2.0 | 0.18 |
| | 4 (350) | 56.0 | 78.4 | 140 | 7.5 | 1.7 | 0.17 |
| | 5 (280) | 77.0 | 90.8 | 117 | 7.1 | 1.9 | 0.22 |
| | 6 (140) | 244.6 | 256.8 | 105 | 8.9 | 3.3 | 0.18 |
| | 7 (140) | 45.4 | 71.5 | 157 | 13.3 | 1.6 | 0.20 |
| | 8 (210) | 219.5 | 195.4 | 89 | 6.1 | 3.8 | 0.18 |
| | 9 (140) | 8.7 | 22.3 | 256 | 21.6 | 2.9 | 0.16 |
| | 10 (280) | 35.6 | 49.4 | 138 | 8.3 | 1.3 | 0.20 |
| | 11 (420) | 122.3 | 107.5 | 87 | 4.3 | 2.9 | 0.22 |
| | 12 (150) | 72.0 | 78.7 | 109 | 8.9 | 3.0 | 0.22 |
| | Average | 88.6 | 128.1 | 144 | 2.7 | 2.1 | 0.19 |
| Vaccinium | 1 (160) | 103.6 | 124.9 | 120 | 9.5 | 2.0 | |
| | 2 ( 80) | 128.8 | 122.6 | 95 | 10.6 | 2.0 | |
| | 3 (160) | 28.0 | 36.9 | 131 | 10.4 | 1.1 | |
| | Average | 78.4 | 107.3 | 136 | 8.7 | 1.7 | |

* The number of observations in parentheses

the greater part of the correlative connections with these four factors. Depending on the year, the correlation coefficient ranged from 0.46 to 0.67. In the analysis of the data about the Vaccinium-type pine forests, where co-dominants were practically missing, and the crown coverage and height of the tree stand differed only slightly, the only factor influencing the productivity was the cover of berry-plants ($r$ = 0.38–0.81, average 0.55).

To predict the productivity for the current year, the following equation may be used (significance level 0.05 for all equation coefficients):

Cladonia–Calluna-type pine forests ($r = 0.68$):
$$y = -141.6 + 1.5x_1 + 17.7x_2 + 57.5x_3$$

Vaccinium-type pine forests ($r = 0.72$):
$$y = -123.8 + 2.0x_1 + 79.0x_3$$

where $y$ = number of berries per m²; $x_1$ = cover of berry-plants, %; $x_2$ = density of co-dominants, points; $x_3$ = productivity points, ocular estimation on the basis of generative organs.

When the yield is predicted on the basis of blossoms, the productivity points must be reduced by 1–2 points, when it is predicted on the basis of the ovaries, by 1 point.

Analyses of variance were performed using the following levels for the factors: cover of berry plants: 0–20, 21–40, 41–60, more than 60%; density of co-dominants: 0–1, 2–3, 4–5 points; crown coverage: 0, 0.10–0.40, 0.41–0.60, more than 0.60; height of the tree stand: 0, 0.1–10, more than 10 m. The factors were analysed in pairs and jointly. In the various models, the correlations of the number of berries with the factors in Cladonia–Calluna-type forests varied within the following limits: correlation with cover of berry-plants 0.21–0.29, with density of co-dominants 0.25–0.37, with crown coverage 0.23–0.34, with year 0.18–0.40. The correlation with the height of the tree stand proved to be non-significant and was not taken into account. In the Vaccinium-type pine forests, the correlation between the number of berries and the cover was 0.55, between the number of berries and the year 0.51. In the latter analysis the data are not absolutely correct, because the number of observations at the different factor levels was not equal, but they give an approximate idea of the relationships within the phytocoenosis.

According to the results of the regression and variance analyses, in the different phytocoenoses the productivity depends on different factors and on their joint influence. The simpler the structure of a phytocoenosis, the smaller is the number of factors influencing the productivity, but the stronger is the correlation with the determining factors and the more exact is the prediction of productivity. In better years, the dependence of productivity on the factors investigated is stronger, but even in the most productive years the crop also depends on several factors which have not been dealt with in this article.

On the whole, the greater the cover of berry-plants (with a few exceptions) and the lower the density of co-dominants, the better is the yield. The optimum crown coverage is 0.10–0.40.

### Stability of the productivity

Although the productivity differs from year to year, the same plots always give more berries than others. To find out the stability of productivity, a two-factor hierarchical variance analysis was carried out, in which "a sample plot" represented one of the factors while "a group of subplots within one sample plot" represented the other factor. The variance ratio between the mean squares of the two factors and the residual deviation for all the study years proved to be significant at the 0.05 level, which means that the productivity in each year is strongly dependent on differences between sample plots and on differences between subplots within a sample plot, the dependence on differences between sample plots being 1.2–2.5 times as great as the dependence on differences between groups of subplots. This means that differences between coenopopulations are greater than those between clones within a coenopopulation. In the different years the correlation between the joint influence of the two factors and the number of berries ranges from 0.57 to 0.74. Most of the sample plots maintained their place in the hierarchy in six years out of seven; only two plots, both with productivity near the average, changed their places. In order to determine the stability of productivity of groups of subplots (clones) within a sample plot, a pair correlation analysis was carried out on model coefficients of the different years calculated on the basis of the analysis of variance. This confirmed that the same plots give better crops every year. It goes without saying that there are some deviations, e.g. in 1983 and in two sample plots in 1978.

### Phenology and the dynamics of crop formation of the cowberry

Phenological investigations were carried out in the Cladonia–Calluna-type pine forests.

It is possible to predict the productivity of the cowberry in the autumn of the preceding year. Inaccuracy is caused by certain difficulties in distinguishing leaf buds and blossom buds. The ratio of blossom buds counted in spring to those counted in autumn is 1.34 ± 0.05. The blossom buds counted in autumn develop 2.21 ± 0.13 berries. The blossom buds counted in spring develop 5.63 ± 0.09 blossoms, these blossoms develop 0.49 ± 0.01 ovaries, and the ovaries develop 0.58 ± 0.01 berries. Thus, the blossoms develop 0.29 ± 0.01 berries, and the blossom buds 1.68 ± 0.05 berries. In 1982 the productivity was low, as only 0.38 (less than the average) of the blossoms developed into ovaries; in 1983 it was low because of the small number of blossom buds which developed the previous year. A blossom flowers during a period of 5–16 days, on the average 8 days, depending on the weather conditions. Self-pollination takes place side by side with insect pollination. The number of ovaries developed in isolated blossoms was 0.44 ± 0.03; the number of ovaries developed in isolated inflorescences averaged 4.00 ± 0.14, which is practically equivalent to fertilization in ordinary conditions. The germination of seeds in the laboratory was 55%. In natural conditions the germination of seeds is also satisfactory, but shoots perish later from short-age of moisture. Sexual reproduction occurs only when seeds fall near tree stumps, fallen trunks or stones.

On the average, the phenology of the cowberry in North-West Estonia is as follows: flowering begins on 3 June, mass flowering on 12 June, flowering ends on 30 June, the berries begin to ripen on 9 August, mass ripening occurs on 24 August and ripening is completed by 3 September.

*Table 2. Sums of mean daily air temperatures necessary for the start of the phenophases.*

| Threshold temperature °C | Flowering Beginning | Flowering Mass | Ripening Beginning | Ripening Mass |
|---|---|---|---|---|
| > 0 | 464 ± 11 | 580 ± 12 | 1448 ± 32 | 1697 ± 27 |
| > 2 | 347 ± 17 | 440 ± 11 | 1200 ± 30 | 1417 ± 27 |
| > 4 | 251 ± 15 | 331 ± 10 | 978 ± 25 | 1167 ± 26 |
| > 5 | 208 ± 13 | 277 ± 9 | 877 ± 31 | 1036 ± 20 |
| > 6 | 170 ± 11 | 231 ± 9 | 765 ± 24 | 911 ± 22 |
| > 8 | 111 ± 9 | 155 ± 11 | 564 ± 15 | 684 ± 13 |
| >10 | 69 ± 5 | 93 ± 10 | 393 ± 15 | 477 ± 12 |
| Period >10 | 238 ± 22 | 354 ± 19 | 1222 ± 29 | 1471 ± 29 |

For each phenophase to occur, a certain sum of mean air temperatures is needed (Table 2). The start of the phenophases can be predicted with the smallest relative error (1.6–2.3%) by summing the mean daily ambient temperatures above 0°C, which for the cowberry may be considered the minimum effective temperature.

## CONCLUSIONS

In an inventory of berry resources, the number of factors influencing the productivity of the cowberry and the strength of their influence varies with the year and the phytocoenosis. If an inventory is based on forest types, it is necessary to include supplementary material for each forest type (equations, tables, etc.).

Productivity differs with the year and region, but a relatively better yield can be obtained from the same coenopopulations and clones within a coenopopulation, which fact must receive attention in selection work.

Phenological and climatic observations carried out in parallel with resource investigations will throw further light on this valuable berry-plant.

Received 22.III.1988

Acta Bot. Fennica 136:37–42, 1988

# Effects of berry production and deblossoming on growth, carbohydrates and nitrogen compounds in Vaccinium myrtillus in north Finland

T. PAKONEN, K. LAINE, P. HAVAS and E. SAARI

Pakonen, T., Laine, K., Havas, P. & Saari, E. 1988: Effects of berry production and deblossoming on growth, carbohydrates and nitrogen compounds in Vaccinium myrtillus in north Finland. — In: Vänninen, I. & Raatikainen, M. (eds.), Proceedings of the Finnish-Soviet symposium on timber forest resources in Jyväskylä, Finland, 25–29 August 1986: 37–42. Acta Bot. Fennica 136. Helsinki. ISSN 0001-5369 ISBN 951-9469-30-3

Biomass measurements and determinations of carbohydrates and nitrogen compounds were carried out on various parts of sterile, fertile and deblossomed plants of *Vaccinium myrtillus* L. gathered from Oulu, Kuusamo and Kilpisjärvi in early August, towards the end of the growing season. Berry production detracted from vegetative growth in that season's new shoots, while deblossoming increased growth. Berry production did not markedly affect the carbohydrate concentration, which in turn increased in the deblossomed plants from Oulu. Total nitrogen compound values were lower in the plants producing berries than in sterile and deblossomed ones.

Key words: Allocation, deblossoming, *Vaccinium myrtillus*, vegetative growth

*T. Pakonen, K. Laine, P. Havas and E. Saari, Department of Botany, University of Oulu, SF-90570 Oulu, Finland*

## INTRODUCTION

The passage from summer to winter poses a mighty ecological challenge for plants in the boreal coniferous zone, including specific needs to reallocate their resources to ensure survival, growth and reproduction. Although reproduction in the *Vaccinium* species takes place vegetatively, they still devote some of their resources to the production of berries and seeds. Some authors claim that increased allocation to seed production reduces vegetative growth (Danilow 1953, Gross 1972, Sohn & Policansky 1977, see also a review by Willson & Burley 1983), and more dry matter and starch has been shown to accumulate in the roots, crowns and leaf blades of deblossomed strawberry plants than in fruiting plants (Forney & Breen 1985).

The aim of the present study was to investigate the effects of berry production and deblossoming on growth, carbohydrates and nitrogen compounds in different parts of bilberry shoots. The plants were deblossomed in June and material representing deblossomed, fertile and sterile plants was collected for chemical analyses and growth measurements in the first week of August. The study is continuing and results presented here are preliminary.

## MATERIAL AND METHODS

Research into the effects of deblossoming and berry production on growth in the dwarf shrub *Vaccinium myrtillus* was commenced in spring 1985 in a forest of the Vaccinium–Myrtillus site type near Oulu (65°N, 25°30'E), a forest of the Hylocomium–Myrtillus site type in Kuusamo (66°20'N, 29°20'E) and in the subalpine birch forest zone at about 500 m a.s.l. at Kilpisjärvi (69°02'N, 20°49'E). Certain plants were deblossomed in June and the branches flowering at that time were marked with coloured rings. From each sampling site 25 deblossomed bilberry plants, 100 fertile ones and 100 sterile ones were then collected in August for measurement and analysis. The samples were separated into 1) current year branches with leaves, 2) previous year's branches and 3) older aerial parts. The branches were measured for length and then dried at 45 °C for 48 hours and weighed. The same material was used for the chemical analyses. The age of the aerial parts was determined from the numbers of annual shoots and microscopically from the annual rings.

For the chemical analyses the samples were ground and passed through a 20-mesh screen. Glucose, fructose, saccharose and starch were assayed by enzymatic methods (Beutler &

Beinstingl 1978) and total nitrogen by the Kjeldahl method. Protein was determined by the Folin-Ciocalteu reagent method (Lowry et al. 1951, Peterson 1977) and soluble low molecular weight nitrogen by the ninhydrin method (Spies 1957), a spectrophotometric procedure based on the chromatographic separation of amino acids and peptides (Moore & Stein 1954).

## RESULTS AND DISCUSSION

Deblossoming increased growth in the plants in terms of both the length and the weight of the shoots, while berry production reduced their length, with differences noticeable even between branches with and without berries on the same fertile plant. The fertile branches were nevertheless heavier than the sterile ones (Fig. 1A and B). Growth declined with the age of the branch, the younger ones allocating more of their resources to this function. The youngest fertile branches were aged three years. Measurements made in Oulu in summer 1985 show a clear connection between the previous summer's growth and berry production, while the fertile branches appear to stem from ones which had grown more vigorously the previous summer than had the sterile branches (Fig. 1C). Evidently a part of the

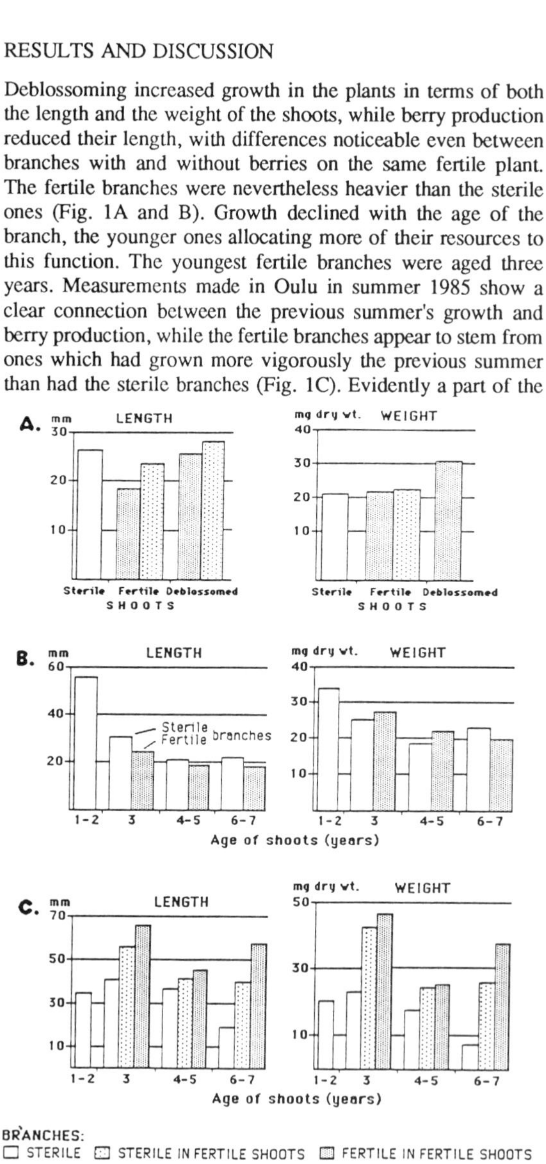


*Fig. 1A–C.* — A. Length (mm) and weight (mg dry wt.) of current-year branches of sterile and fertile shoots of *Vaccinium myrtillus* and of fertile plants after deblossoming. — B. Length (mm) and weight (mg dry wt.) of current-year sterile and fertile branches of *Vaccinium myrtillus* in relation to age. — C. Length (mm) and weight (mg dry wt.) of previous year's sterile and fertile branches of *Vaccinium myrtillus* in relation to current-year branches.

plant which has remained undersized on account of having produced berries or for some other reason will tend to produce sterile branches the following year.

The amount of carbohydrates present is greatest at Kilpisjärvi, the northernmost area studied here, and lowest near Oulu (Fig. 2A), in accordance with the general observation of high carbohydrate values in plants growing in arctic areas (Shaver & Billings 1976). The fact that amounts of photosynthesis products in the bilberry plant do not decline as a consequence of berry production may be indicative of more pronounced photosynthesis in shoots producing berries or greater consumption on vegetative growth in those without berries, as reflected in the growth measurements. A higher level of photosynthesis has certainly been recorded in the strawberry during berry production than during the opening of its buds (Wolkowa et al. 1974). The increase in soluble sugars in the deblossomed shoots compared with the sterile ones was 17% in Oulu and 19% in Kuusamo, but only 7% at Kilpisjärvi, and in Oulu saccharose was to be found only in deblossomed shoots. Reich (1985) reports saccharose concentrations in deblossomed shoots of apple trees to be double those in nondeblossomed ones. In the present case deblossoming did lead to an increase in starch in the aerial parts of the plants gathered from Kilpisjärvi. No measurements were made here of either carbohydrates or growth in the underground parts of the plants, but it may be that removal of the flowers does stimulate growth of the underground shoots as well, and the transport of photosynthesis products into the underground parts.

The concentrations of nitrogen compounds were found to decrease with advancing age (Table 1, Fig. 2B), so that a high nitrogen concentration is indicative of new growth (Wielgolaski & al. 1975). Thus the amounts of nitrogen compounds in the older green parts of the sterile, fertile and deblossomed plants were 30–35% lower than in the current year's growths near Oulu, 10–20% lower in Kuusamo and 30–40% lower at Kilpisjärvi (Table 1). Berry production is reflected in lower values for nitrogen compounds in the various aerial parts of the plant. Although the material from Kuusamo and Kilpisjärvi had a mean of less than 3 berries per stem, the Oulu material could be divided into stems with 1–3 berries and those with more than 3 berries, whereupon nitrogen compounds were observed to decline with increasing numbers of berries, this being particularly noticeable in protein concentrations (Table 1), for the nitrogen present in the aerial parts of the bilberry is chiefly in protein form. Deblossoming renders the concentrations of nitrogen compounds in the bilberry plant virtually identical to those found in sterile plants, although the protein content was higher than in sterile plants at the Kilpisjärvi site. Since the opposite situation prevails in the case of carbohydrates, it may be assumed that deblossomed plants use some of the carbohydrate to synthesize proteins.

The concentrations of soluble low molecular weight nitrogen compounds (compounds possessing $NH_2$ groups, including amino acids, peptides, primary amines and ammonia) vary very little from one part of the plant to another, or between parts of differing ages, by the end of the growing season. Correspondingly, the amino acid nitrogen concentration in birch leaves has been shown to be highest during protein synthesis in the spring and protein hydrolysis in the autumn, while remaining constant throughout the major part of the growing season (Chapin & Kedrowski 1983). In the present case higher concentrations of these compounds were recorded at Kilpisjärvi

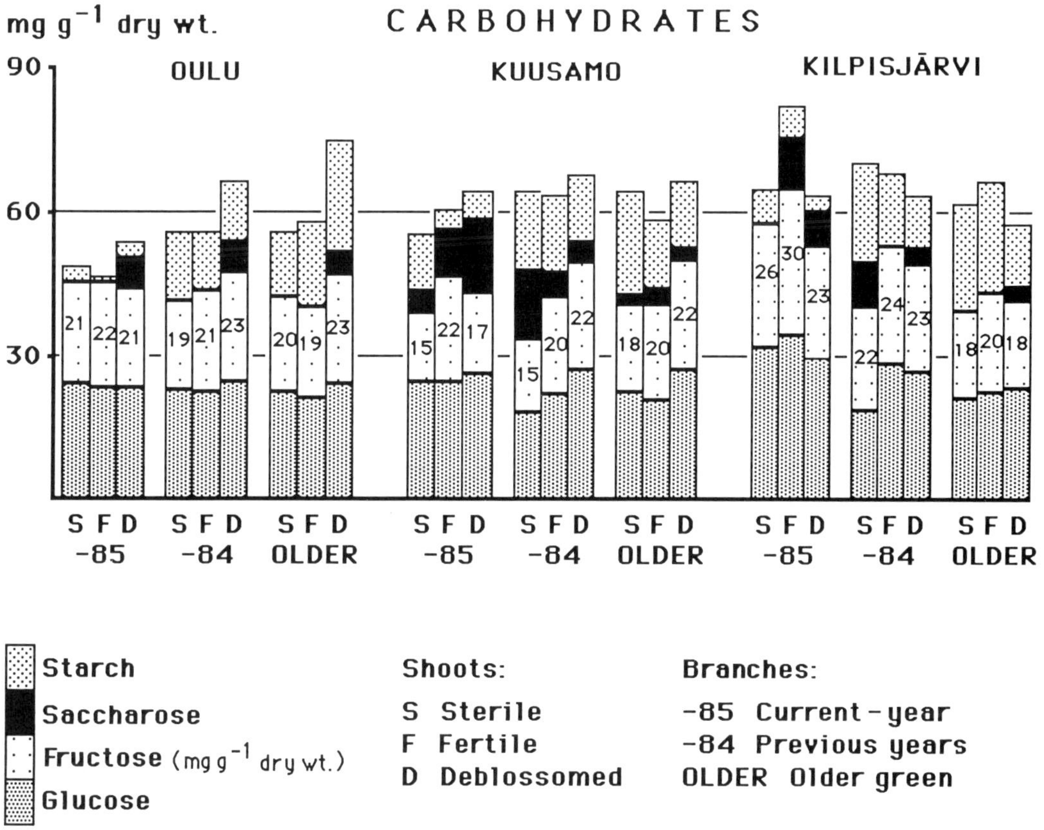


*Fig. 2A.* Carbohydrate concentrations (mg $g^{-1}$ dry wt.) in sterile, fertile and deblossomed shoots of *Vaccinium myrtillus* at three sites in August 1985.

Table 1. *Concentrations of nitrogen compounds (mg $g^{-1}$ dry wt.) in current-year and older branches of the bilberry as measured in August 1985.* (The fertile branches are divided into those with 1–3 berries and those with >3 berries, Sol. N = soluble low molecular weight nitrogen compounds, ·· = not determined).

| | Oulu | | | | Kuusamo | | | | Kilpisjärvi | | | |
|---|---|---|---|---|---|---|---|---|---|---|---|---|
| Branches: | Tot. N | Prot. | Sol. N | Σ | Tot. N | Prot. | Sol. N | Σ | Tot. N | Prot. | Sol. N | Σ |
| Current-year | | | | | | | | | | | | |
| Sterile | 17.0 | 48.0 | 1.7 | 66.7 | 13.9 | 43.6 | 1.7 | 59.2 | 17.6 | 45.5 | 2.7 | 65.8 |
| Fertile, >3 berries | 17.7 | 30.5 | 1.8 | 50.0 | ·· | ·· | ·· | | ·· | ·· | ·· | |
| Fertile, 1-3 berries | 18.1 | 39.3 | 1.9 | 59.3 | 11.5 | 44.2 | 1.4 | 57.1 | 12.7 | 43.7 | 2.9 | 59.3 |
| Deblossomed | 18.0 | 46.0 | 1.5 | 65.5 | 11.2 | 50.9 | 1.6 | 63.7 | 16.8 | 54.2 | 2.9 | 73.9 |
| Older | | | | | | | | | | | | |
| Sterile | 9.5 | 36.2 | 1.1 | 46.8 | 9.2 | 42.0 | 1.2 | 52.4 | 10.8 | 27.1 | 1.3 | 39.2 |
| Fertile, >3 berries | 8.5 | 23.2 | 1.2 | 32.9 | ·· | ·· | ·· | | ·· | ·· | ·· | |
| Fertile, 1-3 berries | 10.3 | 30.9 | 1.1 | 42.3 | 6.8 | 37.1 | 0.9 | 44.8 | 7.2 | 33.4 | 1.4 | 42.0 |
| Deblossomed | 9.5 | 33.1 | 0.9 | 43.5 | 8.2 | 42.0 | 1.1 | 51.3 | 10.5 | 35.0 | 1.1 | 46.6 |

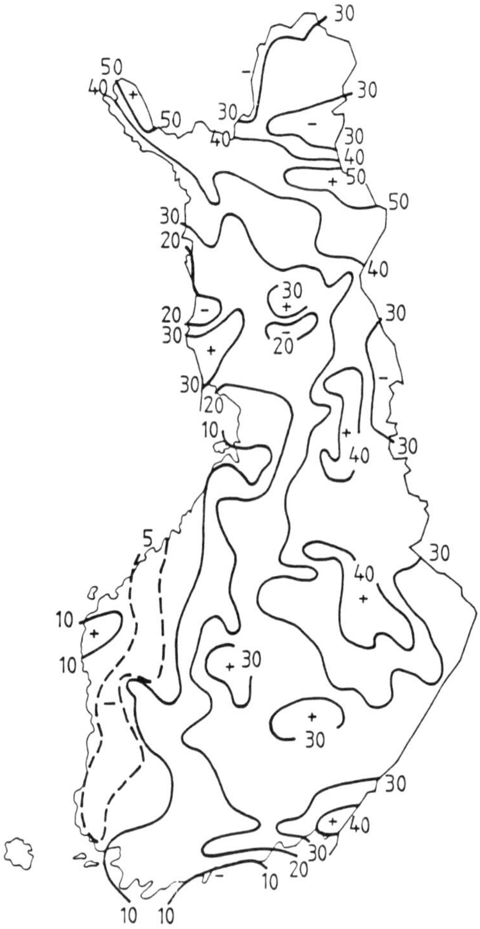


*Fig. 1.* Depth of the snow cover (cm) 15.1.1985. (Ilmatieteen laitos 1985.)

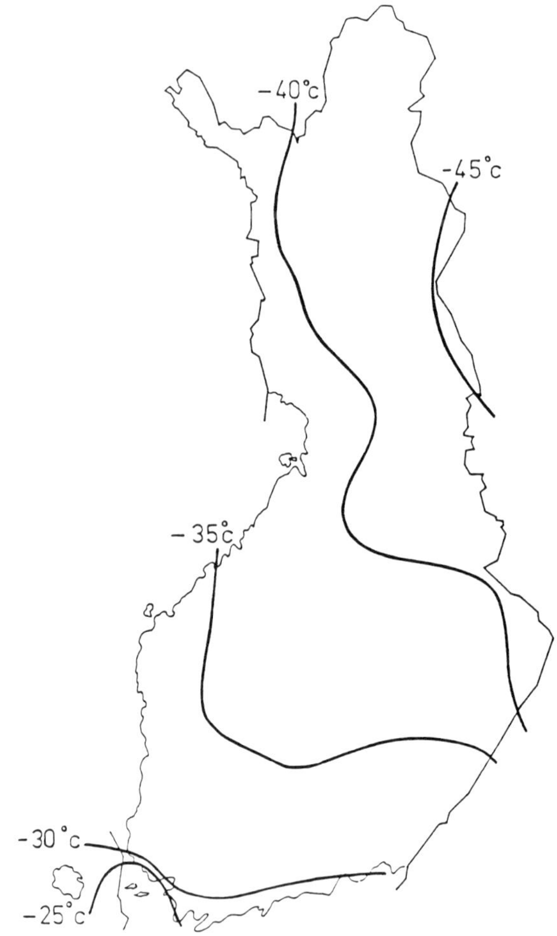


*Fig. 2.* The lowest temperatures in January 1985.

The winter injuries suffered by bilberry buds in different habitats were investigated in Konnevesi (62°30'N, 26°50' E). In May 1985 samples were collected in forests and clearings, under the branches of trees and in sites between trees, and on both flat surfaces and hummocks. Shoot samples were collected in 13 places and their total number was 26. The lowest temperature of the winter, –37°C, was measured at the nearby meteorological station of Jyväskylä on 6 January 1985.

The regional winter injuries sustained by bilberry buds were investigated by collecting shoot samples in southern and central Finland in May and in northern Finland at the beginning of June 1985. Samples were obtained from 58 localities and their total number was 103 (Fig. 3). The samples were taken from berry-picking sites, the collector crossing the berry picking site and cutting a shoot from ground level at every fifth step. Shoots were collected in about 10–30 spots and mailed to the Department of Biology at the University of Jyväskylä.

The damage to lingonberry buds was investigated by collecting 12 samples in 12 municipalities. They were taken in the same way as the bilberry shoot samples and generally also from the same stands, so that they could be easily compared with each other.

The flower buds were examined on each sample, being opened with tweezers under a microscope. Approximately 129 (20–532) buds were studied per sample and classified as damaged and undamaged. Observations were also on the damage to leaf buds.

In Konnevesi damaged and undamaged flower buds were marked in the field and their development was monitored, photographs being taken at different development phases. After the 1985 summer season an inquiry was sent to berry pickers in different parts of Finland. They were asked whether the bilberry and lingonberry crop in 1985 had been very scanty, scanty, moderate, abundant or very abundant, and in what kinds of places the crop was scantiest and most abundant.

## RESULTS

### Winter injuries to bilberry buds

Four degrees of damage were distinguished:

1. The leaf buds surrounding the flower bud are green and undamaged, but the flower bud is partly damaged and the stamens or some of the pistils have turned brown.
2. The leaves surrounding the flower bud are undamaged, but the flower bud has been totally destroyed and turned brown.
3. In addition to the flower bud, the nearest leaf buds protecting it have been damaged and turned brown.
4. The flower bud and all the protecting leaf buds have been damaged and turned brown.

Flower buds which had been damaged during the winter turned brown in spring and gradually softened. They often became detached from the shoot while still inside the gemma and fell to the ground when the bud opened. If the flower bud did not become detached, it dried up as the bud opened and could be seen as a dark pin head in the middle of the leaves. The more severely damaged the bud was, the smaller it remained and the later it opened. Buds which had suffered damage of the first and second degree swelled in the same way as undamaged buds and opened at the same time or somewhat later. Buds which had suffered damage of the third degree opened in nearly the same way as undamaged buds, but noticeably later. The buds with damage of the fourth degree did not open at all. They long remained attached to the shoots, gradually dried up and fell off very easily, leaving a small dent in the shoot. When the greater part of the healthy bilberry buds had opened, the severely damaged shoots could be clearly distinguished by their unopened buds. Sometimes the whole shoot had been destroyed.

### Winter damage to the bilberry in different habitats

In the experiment conducted in Pihtipudas, no snow cover formed on the berry stand during the winter. All the flower buds and even the leaf buds were destroyed, as against only 20% of the 150 flower buds examined in the snow-covered control area.

In Konnevesi on the average 55% of the bilberry flower buds under tree branches were destroyed during the winter, whereas only 19% ($t = 3.60^*$, $df = 6$) were destroyed in the areas between trees. The greater damage was caused by the thinner snow cover under the trees.

The damage to bilberry buds on the hummocks was 43% and on adjoining flat surfaces it was 30%. The difference was probably caused by cold air reaching the bilberry buds on the hummocks through gaps in the uneven snow cover.

In forest clearings 32% of the bilberry buds were destroyed and in nearby forests 35%. The percentages were nearly the same, which was probably due to an equally poor snow cover.

### Winter damage to the bilberry in different areas

The damage percentage of the bilberry buds varied from 3 to 100 (Fig. 3). On the basis of the severity of the damage, Finland was divided into three areas:

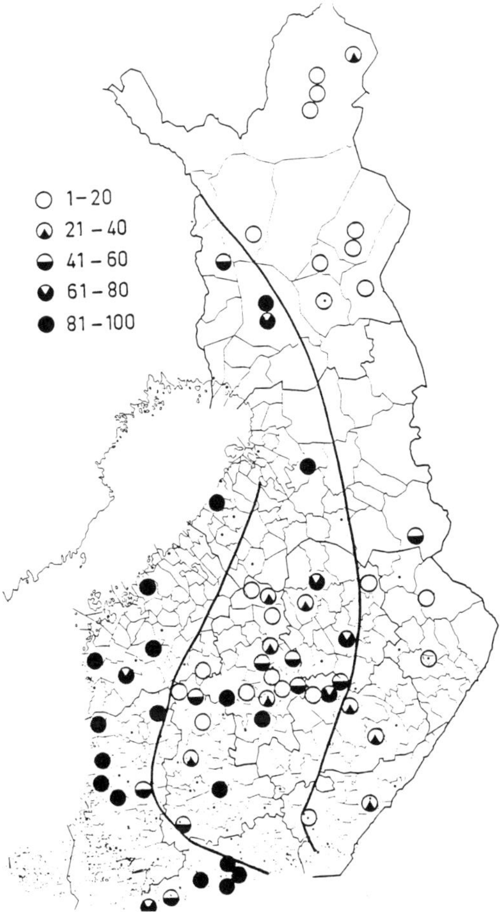


*Fig. 3.* The damage percentages of bilberry flower buds during the 1984–1985 winter season.

I Coastal Finland, where an average of 89% (variation 54–100%) of the buds were destroyed. The damage was great everywhere and local variation was rather small.

II Central Finland, where approximately 46% (5–96%) of the buds were destroyed. The damage percentages varied greatly and were highest in the most southern and northern parts of the area.

III Eastern and northern Finland, where the average damage percentage was 16 (3–46). In this area the damage percentage varied less than in area II.

Damage was clearly most severe when the depth of snow cover was less than 20 cm and slightest when it was more than 30 cm during the intense cold of January (see Fig. 1).

### The bilberry crop in different areas

According to the inquiry regarding summer 1985, the bilberry crop was an average one, but regional differences were great.

On the basis of the findings from the crop inquiry, Finland was divided into seven areas (Fig. 4).

1. Archipelago Finland, where the bilberry crop was average.
2. Varsinais-Suomi and southern Ostrobothnia, where the bilberry crop was very scanty.
3. Uusimaa, Häme and central Ostrobothnia, where the crop was scanty.
4. Central Finland with its surroundings, where the crop varied greatly, but was generally average.
5. Kymenlääni, where the bilberry crop was abundant or at least average.
6. Northern Karelia, where the bilberry crop was very abundant, and
7. Savo, Kainuu, northern Finland and Lapland, where the crop was generally abundant, but varied from average to very abundant.

The bilberry crop was very scanty or scanty in those areas where the snow cover during the cold weather in January was less than 20 cm and the damage suffered by the flower buds was greatest. When the thickness of the snow cover was 20–30 cm, the bud damage and the bilberry crop were average. In eastern and northern Finland, where the snow cover was more than 30 cm thick and bud damage was very small, the bilberry crop was abundant or very abundant.

Archipelago Finland seemed to form an exception. In that area there was hardly any snow cover, but the temperature did not fall below –22 – –25°C. The bilberry type that grows in this area is also probably one that tolerates a severe winter with very little snow.

### The winter damage to the lingonberry and its crop

In the Pihtipudas experiment, in which the lingonberry overwintered without a snow cover, all the flower buds were destroyed.

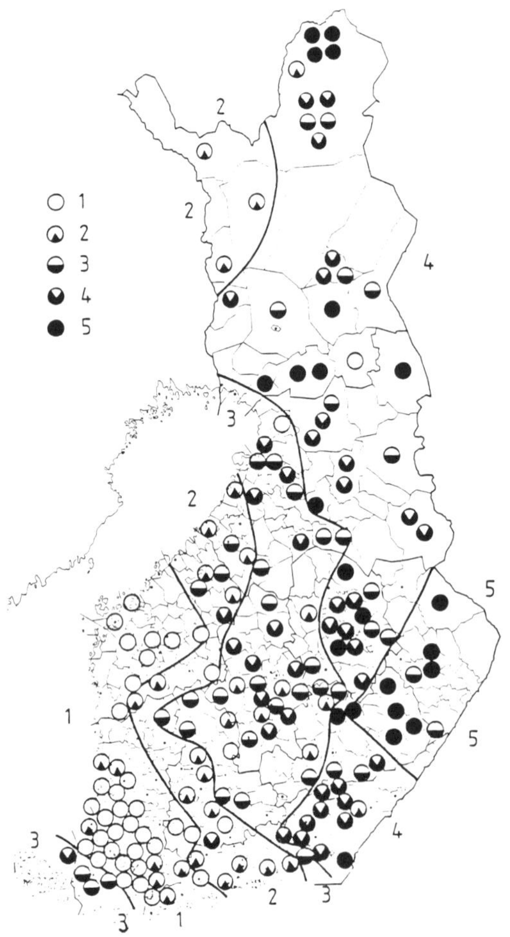


*Fig. 4.* The bilberry crop in 1985. 1 = very scanty, 2 = scanty, 3 = moderate, 4 = abundant and 5 = very abundant.

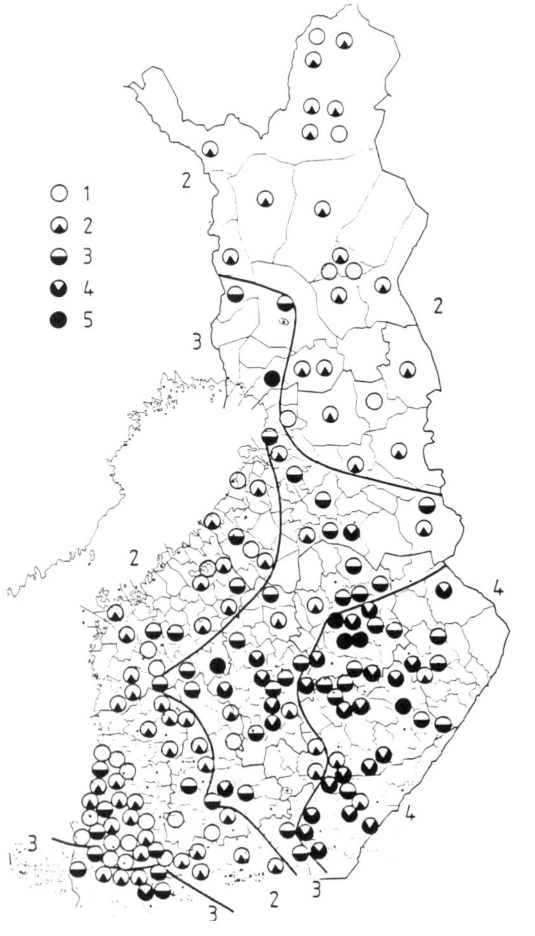


*Fig. 5.* The lingonberry crop in 1985. Symbols as in Fig. 4.

Comparison of bilberry and lingonberry samples collected in different parts of Finland indicated that during winter 1984–1985 the proportion of the bilberry buds that were destroyed was 43.3%, whereas the value for the lingonberry buds was 31%. The difference was significant at the 0.05 level (t-test, $t = 2.02$, $df = 87$).

The following winter was one with abundant snow and the temperature did not fall under 30°C during the period of very little snow. In this winter only 8.5% of the bilberry buds were destroyed and 3.6% of the lingonberry buds. The difference between the species was not significant ($t = 0.21$, $df = 23$).

According to the inquiry conducted on the 1985 crop, the lingonberry crop was scantier than average (Fig. 5). Finland was divided into three areas on the basis of the crop.

1. The region of Turku, where the lingonberry crop was very scanty.
2. Southern, western and northern Finland and Lapland, where the lingonberry crop was scanty or not more than average.
3. Lake Finland, where the lingonberry crop was average, but varied from scanty to abundant.

The lingonberry crop was smallest in the area where the depth of the snow cover was less than 10 cm during the extremely cold days of January. An average crop was obtained in areas where the thickness of the snow cover was 20–30 cm, and an abundant crop in those areas where the snow depth was over 30 cm. An exception was formed by northern Finland and Lapland, where the crop was small, although the depth of the snow exceeded 30 cm. The factors limiting the crop in this case are unknown.

## DISCUSSION

The average height of bilberry shoots varies from 17 to 23 cm and that of lingonberry shoots from 10 to 17 cm. The heights of the shoots are in positive correlation with the canopy coverage of the tree stand (Raatikainen et al. 1984). Hence the depth of the snow cover should be over 25 cm for the bilberry and over 20 cm for the lingonberry, to ensure that most of the flower buds in the top parts of the shoots are covered by snow. In dense forests, a large part of the snow remains in the trees, and tall shoots underneath the trees are left exposed to frosts above the snow cover.

As the bilberry and lingonberry buds could not tolerate frost below –32 – –25°C in January, the greater part of the bilberry buds and a significant part of the lingonberry buds were destroyed in western and southern Finland, which had only a little snow. This was predicted by Raatikainen (1985) in January and by Solantie (1985) in May.

Regional variation was great, however, especially in Central Finland. The best yields were from plants which had overwintered under a thick snow cover and which grew in areas raised above their surroundings (Solantie 1985), by short shoots in sparse forests and by plants growing in areas outside the tree canopy. According to Solantie (1985), the thin snow cover also prevented the warmth conducted by the earth from warming the buds of the berry plants.

Due to the regional variation, nearly total crop failures and excellent bilberry and lingonberry crops occurred locally in Central Finland. Consequently, everyone could acquire the amount of berries needed from the good berry sites, although the crop was only average. However, the great local variation decreased the general interest in berry picking.

## REFERENCES

Havas, P. 1971: The water economy of the bilberry (Vaccinium myrtillus) under winter conditions. — Rep. Kevo Subarctic Stat. 8:41–52.

Ilmatieteen laitos. Kuukausikatsaus Suomen ilmastoon 78, joulukuu 1984, 1–12.

Ilmatieteen laitos. Kuukausikatsaus Suomen ilmastoon, tammikuu 1985, 1–12.

Raatikainen, M. 1984: Syötävien luonnonmarjojemme sadoista ja poiminnasta. — Academia Scientiarum Fennica. Year Book 1982–83: 135–158.

Raatikainen, M. 1985: Pakkanen jättää jälkensä luontoon. — Keskisuomalainen 115(6): 6.

Raatikainen, M. & Pöntinen, M. 1983: Puolukkasadon ennustamisesta. — Metsäntutkimuslaitoksen tiedonantoja 90: 104–115.

Raatikainen, M., Rossi, E., Huovinen, J., Koskela, M.-L., Niemelä, M. & Raatikainen, T. 1984: Metsä- ja suomarjasadot Väli-Suomessa. — Silva Fennica 18:199–219.

Solantie, R. 1980: Kesän 1979 mustikkasato ja edellisen talven pakkaset. — Maataloushallinnon aikakauskirja 10(2): 5–8.

Solantie, R. 1985: Pakkastalven jälkeen. Näin on marjat. — Pellervo 1985(8): 18–19.

Received 28.VII.1987

Acta Bot. Fennica 136:49–59, 1988

# Geographical variation in the phenology and morphology of bilberry in Finland

IRENE VÄNNINEN, SEPPO LAAKSO and MIKKO RAATIKAINEN

Vänninen, I., Laakso, S. & Raatikainen, M. 1988: Geographical variation in the phenology and morphology of bilberry in Finland. — In: Vänninen, I. & Raatikainen, M. (eds.), Proceedings of the Finnish-Soviet symposium on timber forest resources in Jyväskylä, Finland, 25–29 August 1986: 49-59. Acta Bot. Fennica 136. Helsinki. ISSN 0001-5369 ISBN 951-9469-30-3

In 1985 observations were made of the phenology of 76 bilberry (*Vaccinium myrtillus* L.) clones transplanted to Konnevesi, Central Finland (62°37'N, 26°18'E) from different parts of the country. The timing of the phenology was studied in relation to the original latitude and longitude of the clones. Cluster analysis grouped the clones into two ecotypes, northern and southern. The transition zone separating the ecotypes lies roughly along the 64th degree of latitude. The clones retained their inherent growth rhythm and seasonal development even after transplantation to uniform environmental conditions. The effect of latitude was clearly evident in the spring and autumn phenology. Northern clones reacted more slowly to the rise of temperature in the spring, but the autumn phenology proceeded faster in them than in the southern clones. The effect of longitude on the phenology was weak and was evident only among the southern clones. The morphological characters supported the division of the clones into two ecotypes.

Key words: Ecotypes, transplantation, *Vaccinium myrtillus*

*Irene Vänninen, Seppo Laakso and Mikko Raatikainen, University of Jyväskylä, Dept. of Biology, Yliopistonkatu 9, SF-40100 Jyväskylä, Finland*

## INTRODUCTION

In plants adaptation to climatic conditions leads to the formation of genetically different climatic ecotypes. The ecotypes can be morphologically very similar, but differences are often apparent in the physiology or ecology of the plants. An easily recognizable transition zone may exist between the ecotypes or there may be gradual change along a climatic gradient.

In the bilberry latitudinal differences exist in the heat and cold resistance of the plants (Bannister 1970, Havas 1971) and in the water economy of the shoots (Havas 1971). The rate of dark respiration has been shown to differ along an altitudinal gradient (Stewart & Bannister 1973). Some latitudinal differences have also been revealed in the morphology. The shoots of northern bilberries are more xeromorphic than those of southern bilberries (Havas 1977) and there are clear differences in leaf size and stem height between bilberries from Lapland and southern Finland (Purasmaa 1969).

The adaptation of plants to their environment can also be merely phenotypic. In this case modifications are formed. To reveal the nature of variation in the physiology and morphology of plants, they must be transferred to similar environmental conditions. If the differences remain even after the transplantation, conclusions about the existence of climatic ecotypes can be made. The existence and basis of the differences can then be confirmed with studies on the biochemistry and genetics of the plants of different geographical origin (Cooper 1963). These further studies usually reveal the importance of the inherent properties of the plants. Environmental conditions rarely explain the reactions of the plants satisfactorily (Dahl & Mork 1959). Studies on forest trees show that the growth rhythm depends not only on the environmental conditions but on the plants inherent properties as well (Wilhelmi 1957, Lanner 1964, Hari 1968, Hari et al. 1970, Hari & Leikola 1974). The same is true for the annual production of biomass of some dwarf shrubs and herbs (Abrami 1972, Hari et al. 1977). The inherent growth rhythm is genetically coded in the plants and is a specific feature of every individual plant or clone (Walter & Soos 1963, Dietrichson 1967, Hari & Leikola 1974, Andersson 1985). In the bilberry, clones have been shown to differ in the regulation of height growth (Kellomäki 1977).

The aim of this study was to find out whether the bilberries of different geographical origin in Finland can be differentiated into climatic ecotypes on the basis of the timing of phenology in a transplantation experiment. Besides the phenological characters, some morphological properties of the plants were studied as well.

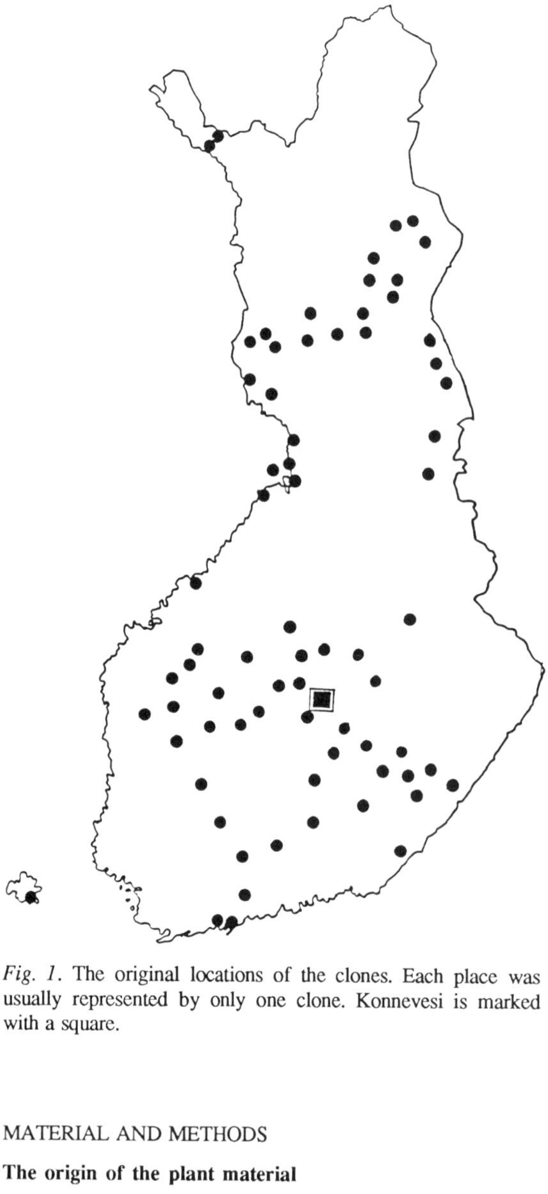

*Fig. 1.* The original locations of the clones. Each place was usually represented by only one clone. Konnevesi is marked with a square.

## MATERIAL AND METHODS

### The origin of the plant material

The material consisted of 76 Finnish bilberry clones (Fig. 1). Most of the clones were collected in spring 1984, but some were taken in the autumn of the same year.

The clones were removed from their original sites with a section of turf 30 x 40 cm in size. They were transplanted some days later at Konnevesi (62°37'N, 26°18'E, about 60 km northeast of Jyväskylä) in an old field, in rows running in an east–west direction. As the field was flat and treeless, each row was sheltered with a wooden fence of a height of 2.3 m. Spaces were left between the boards with which the fences were made, to imitate the light conditions in forest. During rainless periods, the clones were watered. Weeds were removed when necessary.

After the transplantation the most southern clones were growing in a place 300 km to the north of their original habitats and the most northern ones 600 km to the south. This caused noticeable changes in the macroclimate and daylength. The northern clones experienced a lengthening of the growing season by almost one month, whereas for most of the southern clones the change was rather small (from 0 to 10 days). For the clones of northern origin the seasonal accumulated temperature increased by 100–400°C. For the southern clones it decreased by approximately 100–200°C.

For the northern clones the daylength decreased by 4 hours, but the southern clones did not experience so marked a change, their daylength increasing by only 0–1 hours.

In the east–west direction the longest distance between the original and new habitat of the clones was about 300 km.

### The phenology of the bilberry

The data presented by Reuter (1951, 1957) were used for the study of the variation in the phenology in different parts of Finland in field conditions.

At Konnevesi the phenological observations of the clones were based on the two uppermost buds on the stems. Besides the leaf buds, we wanted to include at least 10 flower buds from every clone, but some clones did not produce flower buds at all. The phenological data were obtained from observations of 1 458 buds, of which 415 (28.5%) were flower buds (5.3 per clone, on the average) and 1 043 were leaf buds (13.2 per clone).

The following phenological point events were observed in the clones. The codes presented here are also used further on when the phenological point or segment events are referred to:

| | |
|---|---|
| ONSET | The onset of vegetational development (the point when the leaf buds began to swell distinctly in the spring). |
| LBI | The point when the leaf buds began to open |
| LBII | The point when the leaf buds were completely open |
| FB | The start of budding (the flower buds were considered to have emerged when the stalks of the buds were visible from the side) |
| F | The opening of flowers |
| GB | The beginning of the phase of green berries (the swollen ovaries were regarded as green berries when their diameter was at its greatest in the middle part of the berry) |
| B | The ripening of the berries |
| CG | The cessation of the growth of the current year's shoots |
| BAC | The beginning of autumn colouring (the appearance of the first red or yellow leaves) |

Observations were made daily in the phase of bud and flower opening and when the berries were ripening. Otherwise the interval between observations was 2–4 days. Observations on the dynamics of autumn colouring were made once a week by estimating the percentage of shed leaves or leaves having lost their green colour.

The point when the growth of the current year's shoots ceased was observed in 1985 in 37 clones only. From each clone 4 stems were selected for study. The growth of the uppermost new shoot in the stem was measured by holding it against graph scale paper at intervals of 2 days until it ceased. The measuring was repeated in 1986 on the same stems.

### The morphology of the clones

The morphological characters scored were the length of the current year's shoots, the height of the stems and the leaf index (the ratio of leaf width to leaf length). They were measured from the 37 clones that were used for the study of the cessation of the growth. The length of all the current year's shoots of 4 stems per clone was measured in August 1986. The age of the stems was assessed on the basis of the number of internodes. The height of the stems was measured in 1985. To obtain the leaf index 10 leaves were collected from each clone in August 1986. One leaf per stem in each clone (the 3rd leaf from the top) was used.

### Statistical methods

Cluster analysis was used to differentiate the clones into potential groups. For the analysis the lengths of the following phenophases were converted into accumulated day degrees: ONSET–LBI, LBI–LBII, ONSET–LBII, FB–F, LBII–CG, F–B and GB–B. The actual length (days) of these phenophases could not be used, because the temperature conditions could not be controlled in the field. To improve the reliability of the clustering outcome two different methods were used: the average linkage method and the EML (equal maximum variance likelihood) method.

The variation between different groups of clones was tested using one-way analysis of variance and t-test.

Stepwise multiple regression analysis was used to relate the variation among the phenological and morphological variates to the geographical origin of the clones, represented by latitude and longitude.

The following symbols are used for the statistical significance of the differences: * = $p<0.05$, ** = $p<0.01$, *** = $p<0.001$.

## RESULTS

### The phenology of the bilberry in the field in different parts of Finland

Towards the north, the flowering of the bilberry is delayed by 2.9 days per degree of latitude. In the southernmost part of Finland the bilberry starts to flower on 23 May, on the average, but in northern Finland only on 12 June (Fig. 2). The bilberry does not begin to flower until the mean daily temperature is roughly +10°C.

The timing of the phenology of the plants is usually reversed in the autumn, northern plants setting fruit earlier than southern ones. This is not the case with the bilberry, however, whose berries ripen almost three weeks later in the north than in the south. So the period from the beginning of flowering to ripening of the berries lasts approximately 50 days in both the south and the north. When these days are converted into accumulated day degrees, however, a difference is revealed. The northern bilberries need a smaller sum of day degrees for the berries to ripen (430±14°C, based on observations from 8 different localities) than the southern ones (512±7°C, 17 localities). These values were calculated for the period from the flower opening to ripening of the berries using the dates presented in Fig. 2 and the accumulated temperatures in different localities presented by Heino and Hellsten (1983). The difference between the temperature sums for the north and the south is statistically significant (t-test, $p<0.05$).

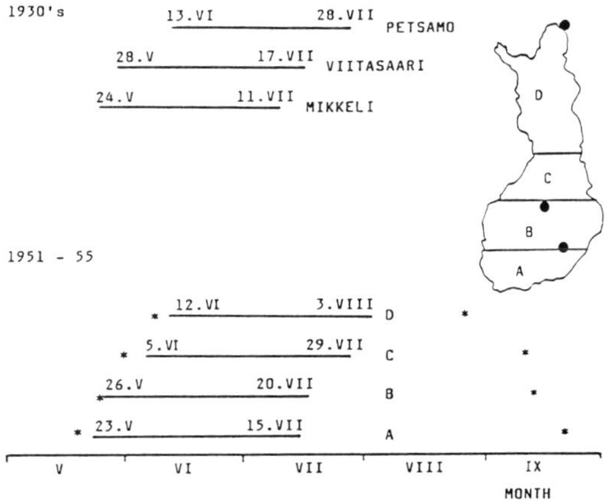


*Fig. 2.* The time of the beginning of the flowering of the bilberry and the time of the ripening of the berries in different parts of the country in the field. The localities where the values from the 1930s are recorded are marked with black dots on the map. The values from the 1950s are for four different zones A–D. The asterisk beside the dates of the 1950s mark the day when the mean daily temperature rose above or fell below +10°C in the zones.

The bilberries in the north have far less time to prepare for winter than their southern counterparts. The time from ripening of the berries to the date when the mean daily temperature falls below +5°C is roughly 55 days in Lapland, 75 days in the Oulu region and 100 days in the southern part of Finland.

### The average phenology of the transplanted clones

In 1985 the buds began to swell after the warm day on 13 May, when the temperature exceeded +12°C. The next 7 days were cold and rainy, however, so that the buds began to enlarge rapidly only during the last week of May. On 26 May the mean daily temperature rose permanently above +10°C. Warm, occasionally even hot weather set in for a week. On 1 June the leaf buds of the clones had opened completely. The flowers opened 2 days later.

The berries ripened on 21 July, on the average (Table 1). Table 2 presents the accumulated temperatures of phenological events.

On 19 October all the clones had either shed all their leaves or the leaves had changed colour from green to red, yellow or brown. The active period of the clones lasted approximately 5 months (from 16 May to 19 October).

Table 1. *The means, standard deviations and ranges of variation of the phenological point events in the clones in 1985.* The range of variation gives the interval between the day when a given event was recorded for the first time in the clones and the day when 90% of the observation units (buds, flowers, berries, clones) had reached the point in question. The days are the number of days elapsed since the beginning of the year. The actual dates are shown in parenthesis.

| Phenological point event | n | Average no. of days | S.D. | Range of variation |
|---|---|---|---|---|
| Beginning of vegetational development | 1178 | 136.3 (16.V.) | 4.52 | 133–139 (13.–19.V.) |
| Leaf buds begin to open | 1202 | 147.6 (28.V.) | 5.55 | 139–152 (19.V.–1.VI.) |
| Leaf buds completely open | 1234 | 152.3 (1.VI.) | 5.22 | 139–158 (19.V.–7.VI.) |
| Start of budding | 341 | 150.6 (31.V.) | 3.66 | 141–152 (21.V.–1.VI.) |
| Opening of flowers | 369 | 153.4 (2.VI.) | 4.52 | 149–158 (29.V.–7.VI.) |
| Cessation of growth | 651 | 176.3 (25.VI.) | 3.71 | 171–182 (20.VI.–1.VII.) |
| Beginning of phase of green berries | 286 | 170.8 (20.VI.) | 4.56 | 164–175 (13.–24.VI.) |
| Ripening of berries | 268 | 202.1 (21.VII.) | 7.15 | 191–212 (10.–31.VII.) |
| Beginning of autumn colouring | 74 | 222.6 (10.VIII.) | 5.25 | 220–233 (8.–21.VIII.) |
| End of autumn colouring | 74 | 292.2 (19.X.) | 9.04 | 283–310 (9.X.–5.XI.) |

Table 2. *The accumulated temperature (°C) at the beginning (B), on the average day (AVE) and at the end (E) on the phenological point events.* The temperatures were taken from the records of the meteorological station at Jyväskylä airport by courtesy of Ilmatieteen laitos.

| Phenological point event | Accumulated temperature | | |
|---|---|---|---|
| | B | AVE | END |
| Beginning of vegetational development | 22.4 | 28.4 | 32.3 |
| Leaf buds begin to open | 32.3 | 71.0 | 133.5 |
| Leaf buds completely open | 32.3 | 115.7 | 133.5 |
| Start of budding | 35.0 | 105.9 | 117.5 |
| Opening of flowers | 85.6 | 121.5 | 133.5 |
| Cessation of growth | 258.1 | 333.5 | 397.1 |
| Beginning of phase of green berries | 177.9 | 258.1 | 313.4 |
| Ripening of berries | 451.5 | 579.4 | 675.6 |

### Clustering of the clones into groups

The two methods used gave similar results, though the details of the grouping of the clones were somewhat different. The average linkage method grouped almost all the clones of southern origin together in one cluster (black dots in Fig. 3A). Among the clones of northern origin this method gave results suggesting that they could be divided still further into two subgroups (white and black squares in Fig. 3A). The clustering obtained with the EML-method suggested, for its part, the division of the clones of southern origin into two subgroups (white circles and black dots in Fig. 3B). There were also some clones of northern origin that the methods used placed in a cluster consisting mainly of clones of southern origin and vice versa. On the basis of the results of the clustering, however, the clones were divided into two subgroups only, the northern and the southern, regarding the 64th degree of latitude as the border between the two groups. The decision is somewhat arbitrary, because there were unfortunately hardly any clones originating from between the 64th and 65th degrees of latitude (Fig. 1).

### The phenology of the clones of different origin

The northern clones started their vegetational development later and remained behind the clones of southern origin in almost every scored phenological character (Fig. 4). This was also revealed in the temperature sums required by the clones for moving through the different phenophases (Table 3). The first phenological phase, ONSET–LBII, differed most significantly between the clones of different origin as far as the accumulated temperature is concerned. The leaf buds of northern clones opened 1.3 days later and required a 13°C higher temperature sum to open than the buds of southern clones. The difference was, however, significant only on the first half of this period (the phase ONSET–LBI in Table 3). When the temperature rose steeply on 26 May the southern clones were the first to react: their leaf buds began to open almost immediately. It took 2 more days for the northern clones to react to the rise of temperature (Fig. 5). From then on, however, the opening of leaf buds was similar in the two clone groups as far as regards the temperature sum required for the complete opening of the buds (45.4°C in both clones, approximately; see Table 3).

Due to the later opening of leaf buds in northern clones their flower buds also appeared later. The weather was very warm during this period and it took only 2–3 days, on the average, for the flowers to open. When counted in days this phase lasted somewhat longer in the northern clones because the mean daily temperature during the period (1–4 June) was 9.4°C, that is to say, nearly 5°C less than the corresponding temperature for the southern clones moving through the same phenophase a few days earlier. In the southern clones this phenophase occurred from 30 May to 2 June. The mean daily temperature during this time was 14°C (Fig. 5). The accumulated

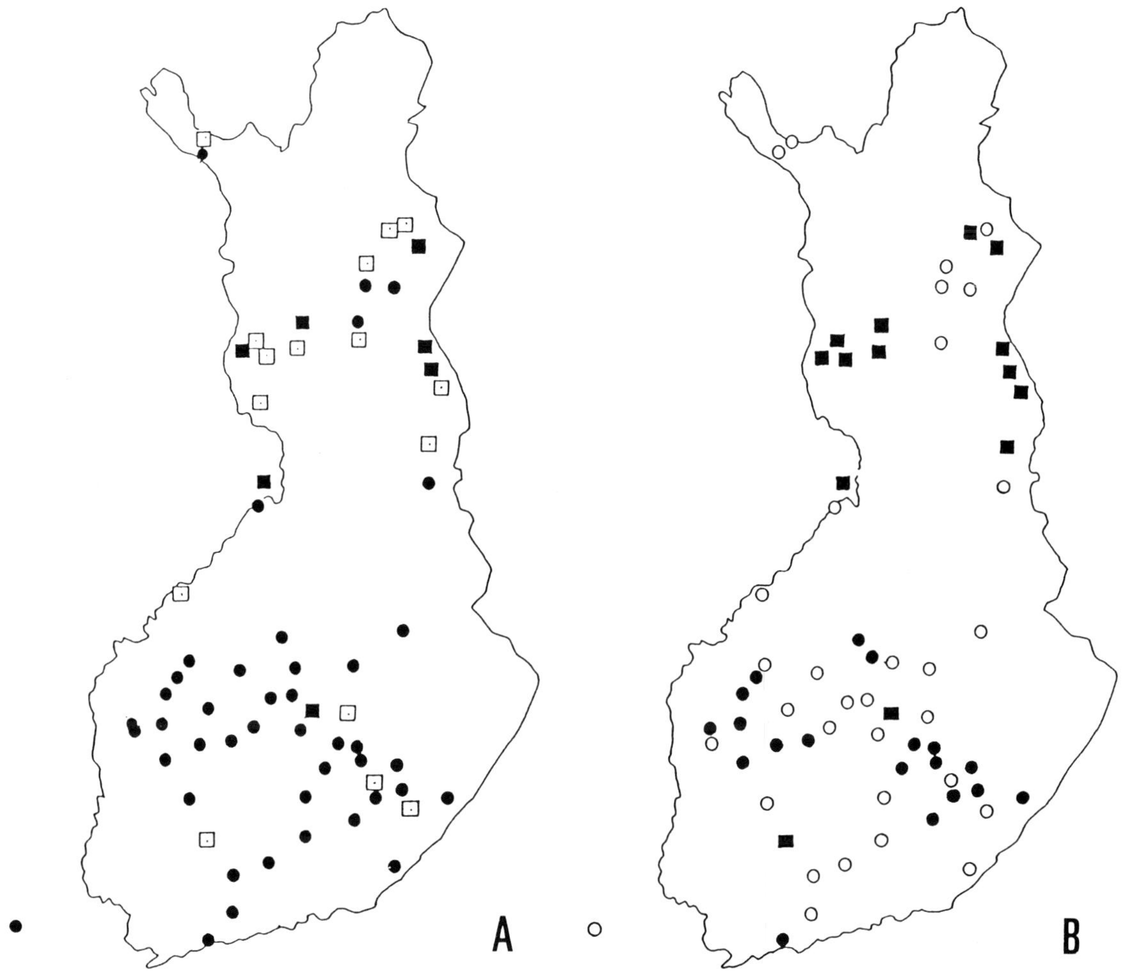


*Fig. 3.* Grouping of the bilberry clones based on the results of the cluster analysis using two different methods. A) the average linkage method B) the EML (equal maximum variance likelihood) method. For explanations, see the text.

temperature required for the flowers to open was similar in the two groups (37.1°C on the average).

The berries of the northern clones ripened somewhat later than those of the southern clones. Here the time from the opening of the flowers to the point when berries turned blue is referred to as the ripening of berries. The time when the flowers were pollinated is not known. The differences were not big, but significant. The northern berries ripened in 51.5 days, the southern in 48.4 days (t-text, $p<0.01$). The difference was also significant when this phenophase was converted into the accumulated temperature (Table 3). The result was the same when the lengths of the phases of green berries (the phase GB–B) in the northern and southern clones were compared.

In 1985 the growth of new green parts ceased on 25 June in southern clones and only one day later in the northern ones. The accumulated temperature for the period from the opening of the leaf buds to the cessation of growth was approximately 214°C in both groups of clones and the total accumulated temperature at that time was 333°C. The next year, however, differences appeared between the northern and southern clones in this respect. The northern clones ceased their growth on 9 June, the southern ones on 12 June. The total accumulated temperatures on these days were 257 and 293°C, respectively, the difference being statistically significant (t-text, $p<0.01$).

In the second half of the growing season, the relationship between the northern and southern clones was reversed in 1985. After the berries had ripened it took about 18 days in the northern clones and 21 days in the southern clones before the first signs of autumn colouring appeared. The autumn colouring began on 10 August, on the average, in both clone groups. The dynamics of autumn colouring and leaf shedding in both groups of clones proceeded similarly as regards the beginning and the end of this phase (Fig. 6). The difference between the clones of northern and southern origin was revealed during the first two

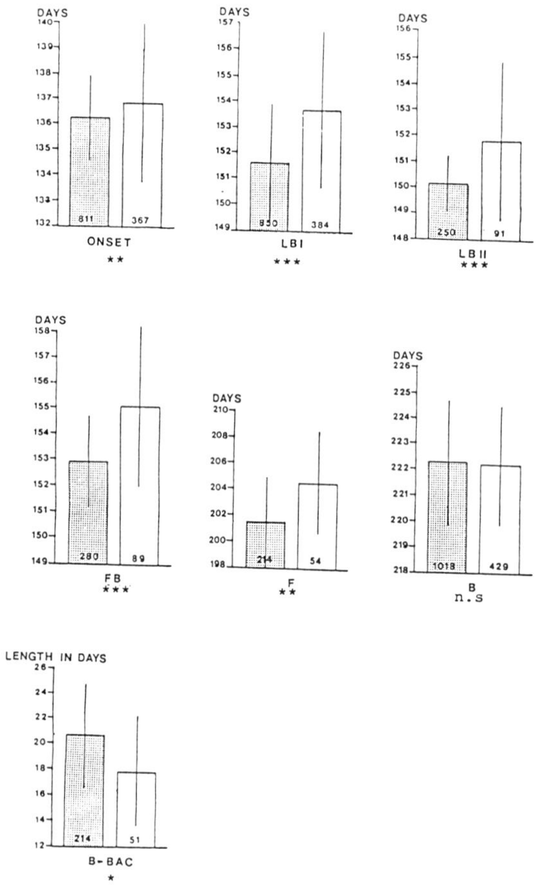


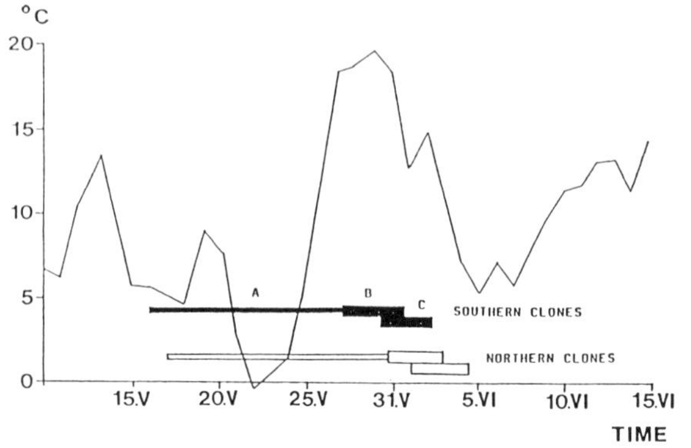


*Fig. 5.* (Above) The first phenophases in spring 1985 in the southern and northern bilberry clones related to the mean daily temperature curve. A = ONSET–LBI, B = LBI–LBII, C = opening of the flowers.

*Fig. 4.* (Left) The beginning, length and end of different phenological point and segment events in clones of southern (dark columns) and northern (white columns) origin. The vertical bars denote the standard deviations. Figures inside the columns refer to the number of cases. 'Days' means the number of days elapsed since the beginning of the year. For the codes of the phenophases see the text.

weeks of October, when leaf shedding intensified markedly. The percentage of shed leaves was significantly higher in the northern clones on 2 and 10 October (t-test, $p<0.031$ and $p<0.01$, respectively) (Fig. 6). On 2 October the northern clones had shed altogether 35% of their leaves. The southern clones reached this stage not until a week later; at that time the northern clones had lost as much as 52% of their leaves.

Table 3. *The accumulated temperature (°C) required by the clones of northern and southern origin to pass through different phenological phases and the significance of the difference between the two clone groups.*

| Phenophase | n | Accumulated temperature °C | S.D. | Significance |
|---|---|---|---|---|
| ONSET–LBI: | | | | |
| Mean | | 45.9 | 35.06 | |
| South | 768 | 42.1 | 33.44 | |
| North | 368 | 53.8 | 37.03 | *** |
| ONSET–LBII: | | | | |
| Mean | | 81.6 | 34.79 | |
| South | 780 | 77.5 | 28.96 | |
| North | 367 | 90.5 | 43.47 | *** |
| F–B: | | | | |
| Mean | | 467.6 | 64.61 | |
| South | 181 | 462.7 | 60.59 | |
| North | 55 | 483.8 | 74.72 | * |
| GB–B: | | | | |
| Mean | | 350.4 | 69.27 | |
| South | 207 | 345.4 | 66.26 | |
| North | 57 | 368.5 | 77.16 | * |

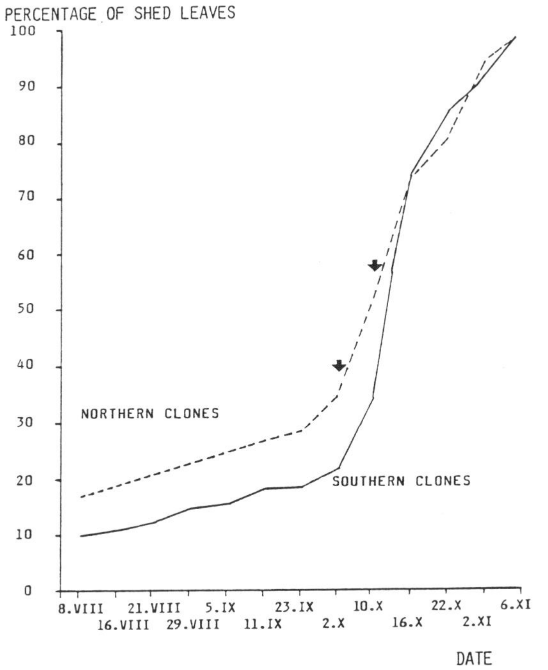


*Fig. 6.* The dynamics of leaf shedding in northern and southern clones (percentage of shed leaves). On October 2 and October 10 (arrows) the clones differed statistically significantly from each other (t-test, $p<0.031$ and $p<0.01$, respectively).

## Morphological characters of the clones

Two of the morphological characters scored were such that had the time to react to the change of climatic conditions after the transplantation, namely the growth of current year's shoots and the leaf index. The third character, stem height, reflected purely the conditions of the original geographical location of the clones.

There was no significant difference in the leaf index between the two groups of clones, whereas both the stem height and the growth of current shoots was significantly less in clones of northern origin (Fig. 7). Both the stem height and the growth of current shoots correlated with the age of the stems. The correlation between the stem height and the age of the stem was 0.39 ($p<0.001$, $n = 94$) and between the growth of current shoots and the age of the stem –0.15 ($p<0.001$, $n = 602$). Therefore the effect of the age was removed with covariate analysis. Age explained 2.2% ($p<0.001$) of the variation in the growth of the current shoots and 15% ($p<0.001$) of that of the stem height. The geographic origin of the clones accounted for 27.4% ($p<0.05$) of the variation in the stem height and 3.1% ($p<0.001$) of that in the growth of the current shoots after the removal of the effect of age.

Although the clones of different origin differed significantly in most of the phenological and morphological characters scored the variation within groups was high. Between groups variation was highest (8.9%) in stem height. For the growth of the current shoots the between groups variation was 2.0%. In phenological characters expressed as cumulative heat sums the between groups variation accounted for 1.9–2.6% of the total variation.

## The relative importance of latitude and longitude on the phenology and morphology of the clones

The regression analysis was made using both pooled data and analysing data from southern and northern clones separately (Tables 4 and 5). Accumulated day-degrees of different phenophases were used in analysing the phenological data. In pooled material, the latitude had the greatest effect on the phenology of the very beginning of the growing season. The effect of longitude began to show in later phases of phenology. When northern clones were treated separately the longitude did not

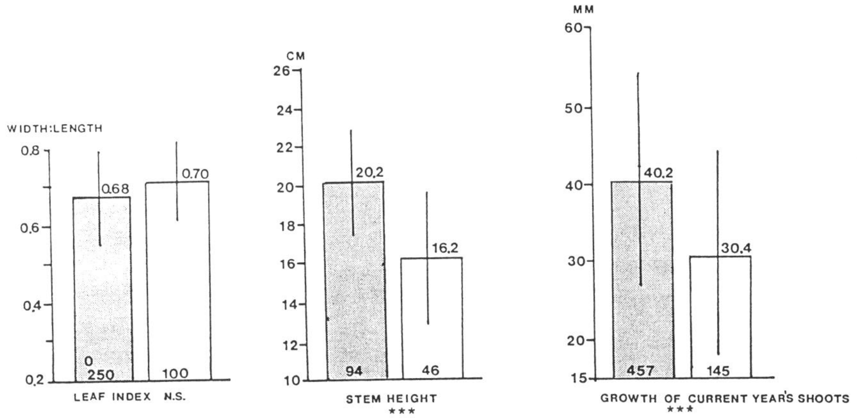


*Fig. 7.* Leaf index (width-to-length ratio), stem height (cm) and growth of the current year's shoots (mm) in the southern (dark columns) and northern (white columns) clones. The vertical bars denote the standard deviation. Figures inside the columns refer to the number of cases, those above the columns denote the mean value in each of the groups.

Table 4. *The regressions of the phenological and morphological characters on latitude (LAT) and longitude (LONG) (in the two morphological characters also on AGE of the stem).* HEIGHT = height of the stems, GR86 = length of the current shoots in 1986.

| Phenological event | Significance of regression on LAT | LONG | Regression equation | Percentage of variation explained by regression |
|---|---|---|---|---|
| ONSET–LBI | *** | – | 3.018 LAT + 145.222 | 4.0 |
| ONSET–LBII | *** | – | 2.918 LAT – 103.084 | 3.8 |
| GB–B | – | ** | 5.384 LONG + 212.878 | 2.9 |
| LBII–CG | – | * | 1.795 LONG + 169.097 | 0.8 |
| B–BAC | – | * | –0.481 LONG + 32.307 | 1.6 |

| Morphological character | LAT | LONG | AGE | | |
|---|---|---|---|---|---|
| HEIGHT | *** | * | *** | –1.095 LAT + 1.783 AGE + 0.474 LONG + 67.309 | 32.8 |
| GR86 | *** | – | ** | –1.801 LAT – 2.296 AGE + 164.604 | 4.6 |

Table 5. *The regressions of the phenological and morphological characters of the northern and southern clones on latitude (LAT) and longitude (LONG) and on AGE (for morphological characters).* For explanations see Table 4.

*Northern clones:*

| Phenological event | Significance of the regression on LAT | LONG | Regression equation |
|---|---|---|---|
| LBI–LBII | * | – | –2.882 LAT + 238.514 |
| FB–F | – | ** | 3.469 LONG – 53.864 |
| LBII–CG | – | * | 4.732 LONG + 89.247 |

| Morphological character | LAT | LONG | AGE | |
|---|---|---|---|---|
| HEIGHT | ** | – | * | –3.005 LAT + 1.699 AGE + 207.513 |

*Southern clones:*

| Phenological event | Significance of regression on LAT | LONG | Regression equation |
|---|---|---|---|
| ONSET–LBI | *** | – | 6.204 LAT – 341.598 |
| ONSET–LBII | *** | – | 5.180 LAT – 242.847 |
| F–B | – | ** | 5.502 LONG + 324.674 |
| GB–B | – | * | 6.776 LONG + 174.902 |
| LBII–CG | ** | *** | –9.719 LAT + 2.664 LONG + 770.448 |

| Morphological character | LAT | LONG | AGE | |
|---|---|---|---|---|
| HEIGHT | – | – | *** | 2.155 AGE + 9.549 |
| GR86 | – | – | ** | –2.188 AGE + 52.896 |

have very much effect on the phenology, whereas in the southern clones the effect of longitudinal origin of the clones was clearly revealed. There was no trend of the height of the stems of becoming shorter towards the north among southern clones, but the age of the stems affected the height and growth of current year's shoots significantly. In northern clones age was of minor importance in explaining the height of the stems, the latitudinal origin of the clones being more important.

## DISCUSSION

### Phenological events of the beginning of the growing season

The bilberry clones of different geographical origin proved to be genetically different. Both the phenological and morphological characters under study supported the separation of southern and northern ecotypes. The phenological differences were clearest for the events at the beginning and end of the growing season. This accords with the conclusion that in climatic conditions with clear seasonal variations the most important factors for plant growth and reproduction are temperature and time (Sarvas 1974). The importance of time is accentuated in the north, where the growing season is short.

The timing of phenology in actual days did not differ very much between the southern and northern ecotypes. This showed the bilberry to be an extremely plastic species which can rapidly match its rhythm of seasonal development to new environmental conditions even without genetic changes. Possible regional differences in phenology can therefore easily be hidden behind the bilberry's sensitivity to weather (Havas 1971).

The progress of the spring development of bilberry stands in the field in different parts of Finland accords with the advance of the +5°C isotherm across the country. This isotherm shows a delay of 3 days per degree of latitude from the south to the north in the spring (Reuter 1951). Solantie (1981) reports that bilberries in the north of Finland begin to flower at a temperature sum 15°C lower than in the south of Finland. The mean daily temperature at the time when the bilberry begins to flower is +10.2°C in southern Finland and +9.2°C in northern Finland (Solantie 1983, cf. also Snigirev & Raptunovich 1980). In view of these observations the northern clones could have been expected to start their development in lower temperatures than the southern ones in the transplantation experiment.

The most probable explanation of the slower reaction of the northern clones to the rise of temperature in the spring is that the northern and southern bilberries were at a different stage of physiological development when the snow melted and the weather became warmer. The shoots of southern bilberries are known to experience a steep rise in the osmotic value in the spring (Havas 1971), which makes them resistant against the cold and dry periods that can occur after the snow has melted. The shoots of northern bilberries show a rise in the osmotic value, too, but it is not so steep and in the bilberries of northern Lapland may be totally lacking. Besides this, the more northern the origin of the plants, the later occurs the peak in the osmotic value. At the time when the snow melted at Konnevesi the southern clones were ready for growth, whereas the rise of temperature came too early for the northern clones and they were not able to respond to it as quickly. Their normal seasonal rhythm was preserved after the transplantation to new climatic conditions. This points to the existence of genetic differences in the inherent growth rhythm between the southern and northern bilberry ecotypes. In addition, it must be noted that not all species show the tendency for northern individuals to be the first to start growth when they are transplanted to lower latitudes (see Nienstaedt 1974 and references therein). Nienstaedt also points out that the change in the photoperiod can modify the heat requirements of the plants in the spring in transplantation experiments.

The growth of the bilberries ceased earlier in our experiment than was reported by, for example, Havas and Kubin (1983) from northern Finland. The earlier cessation of growth in the northern than in the southern clones in 1986 could be explained by the warmer than average weather in June. The mean daily temperature was +3°C higher than normal (Ilmatieteen laitos 1986). The difference in the growth rhythm of southern and northern clones was revealed only when the temperatures were higher than average, that is only in 1986. For the northern clones the weather in June may have been exceptionally warm, while for the southern clones the situation was more normal. The optimal temperatures for the growth of plants of northern origin are known to differ from those for southern ones (Raatikainen et al. 1984). The light quality may also play some role. It is known to affect the duration of growth of *Betula pubescens* of different latitudinal and altitudinal origin (Håbjørg 1972b).

### Phenological events of the middle and of the end of the growing season

The events of the middle of the growing season were rather insensitive to the latitude effect. The effect of longitude was revealed among the southern clones. In the north of Finland both the moisture and continentality gradients run largely parallel to the degrees of latitude, whereas in southern Finland they cut across the latitude (Tuhkanen 1980). This explains why the effect of longitude was revealed mainly among the southern clones.

The northern clones needed a higher sum of day degrees for the berries to ripen, which is contrary to the situation in the field. The changed daylength could explain this contradiction. In the north the plants can take advantage of the continuous illumination in the summer, a fact that compensates for the lower temperature (cf. Håbjørg 1972a). At Konnevesi the maximum daylength was 'only' 20 hours and the higher temperature could not compensate the northern clones for the absence of continuous illumination.

The temperature factor probably interacted with the daylength factor in the autumn as well when autumn colouring began. The northern clones were not the first to show autumn colouring, as could have been expected (the autumn colouring began earlier in the northern clones in the sense that the time from ripening of the berries to the beginning of autumn colouring was shorter in them). The daylength is often considered important for the beginning of autumn colouring (Stern & Roche 1974), but if the temperature is high, the effect of the daylength is not necessarily strong enough to cause the autumn colouring to begin (Håbjørg 1972a, Koski & Selkäinaho 1982). After the beginning of the autumn colouring the biochemical processes involved in this stage in winter hardening proceeded faster in the northern clones and the period of most intensive leaf shedding began earlier than in the southern clones. According to Havas (1971), the osmotic values of bilberry shoots from Lapland rise earlier in the autumn and more sharply than the values of shoots from Oulu and southern Finland.

**Variation in morphological parameters**

The morphology of the bilberry has been shown to vary along an altitudinal gradient. In high altitudes low bilberry stems have a competitive advantage, because this leads to greater rates of photosynthesis through the maximisation of stomatal conductance and the nitrogen economy of the leaves (Woodward 1986). Correlation of the berry yield, stem height, leaf size and berry weight with the altitude was revealed also in a study of bilberry by Moisejeva and Moisejev (1977). Stem height can be thought to be adapted to the depth of snow cover. The average depth of the snow cover increases towards the east and the north. Correspondingly the stem height of the bilberry increased towards the east. The effect of latitude, however, was much more significant. Since the stem showed no tendency to become lower towards the north within the southern clone group, the difference between the northern and southern clones is accentuated. In the north of Finland the snow cover is thick, but the winter temperatures may be so low that low stems are necessary for the plant to remain sufficiently sheltered under the snow cover. The sparseness of the forests in the north may also contribute to the shortness of the stems in Lapland. Bilberry stems are known to be short in habitats with abundant light (Raatikainen et al. 1984). The phenomenon of plants becoming lower towards the north is seen in many plant species in Scandinavia.

The observations of the annual growth of the bilberry in different areas agree with those of Håbjørg (1972a, b), who studied the reactions of *Betula pubescens* of different latitudinal origin to daylength, temperature, light quality and light intensity. The annual growth of plants from the farthest north (70°N, Alta, Norway) was less than the growth of more southern individuals (63°N in Trondheim, Norway and 56°N in Århus, Denmark). He also noted that the differences between plants from 63°N and 56°N was not significant, but that the most northern specimens (70°N) differed clearly from the plants of the two southern localities. This proved to be the case with the bilberry, too. A tendency for the annual growth to become smaller towards the north was not evident within the southern or northern clone groups treated separately, but a clear difference was found when the southern and northern groups were compared with each other.

**The steepness of the latitudinal and longitudinal climatic gradient and their effect on differentiation of ecotypes in the bilberry**

Temperature and daylength can be regarded as the most important factors causing differentiation of climatic ecotypes in the bilberry on the south–north axis. The situation is, however, complicated by the fact that the continentality factor affects the differentiation of climatic ecotypes on the west–east axis as well, especially in the south. The classification of the bilberry into geographically clearly different ecotypes on the west–east axis is difficult. The borders between climatic ecotypes are never as distinct as between edaphic ecotypes (Turesson 1930). The change of temperature and light conditions towards the east over a distance of 400–500 km is not as radical as the change towards the north over a distance of more than 1 000 km. The variation in the phenology and morphology of the bilberry along the longitudinal gradient is consequently gradual and no distinct climatic ecotypes have developed. They could become evident, however, if we travelled farther east and were able to compare bilberries of more continental origin with those growing in the Finnish territory.

REFERENCES


Abrami, G. 1972: Optimum mean temperature for plant growth calculated by a new method of summation. — Ecology 53:893–900.

Andersson, B. 1985: Frystest av norrländska plantageskördar. Freezing test of northern Scots pine seedlings from seed orchards. — In: Nilsson, J.-E. (ed.), Skogsodlingsmaterial för hårda nordliga lägen. Föredrag från ett seminarium i Umeå 19–20 februari 1985. Swedish University of Agricultural Sciences, Dept. of Forest Genetics and plant physiology, Rapport 4:42–56. Umeå.

Bannister, P. 1970: The annual course of drought and heat resistance in heath plants from an oceanic environment. — Flora (Jena) 159:105–123.

Cooper, J.P. 1963: Species and population differences in climatic response. — In: Evans, L.T. (ed.), Environmental control of plant growth: 381–404.

Dahl, E. & Mork, E. 1959: Om sambandet mellan temperatur, andning og vekst hos gran (Picea abies (L.) Karst.). Summary: On the relationship between temperature, respiration and growth in Norway spruce (Picea abies (L.) Karst.). — Meddel. Norske Skogforsoksv. 53:81–93.

Dietrichson, J. 1967: Broad sense heritability estimates of growth rhythm and height growth of Norway spruce (Picea abies (L.) Karst.) seedlings of southern Norwegian origin. — Meddel. Norske Skogforsoksv. 23:201–221.

Håbjørg, A. 1972a: Effects of photoperiod and temperature on growth and development of three latitudinal and three altitudinal populations of Betula pubescens Ehrh. — Meld. fr. Norges Landbrukshogskole 51(2): 1–26.

Håbjørg, A. 1972b: Effects of light quality, light intensity and night temperature on growth and development of three latitudinal populations of Betula pubescens Ehrh. — Meld. fr. Norges Landbrukshogskole 51(26): 1–17.

Hari, P. 1968: A growth model for biological population, applied to a stand of pine. — Commun. Inst. Forestalis Fenniae 66(7): 1–16.

Hari, P., Kellomäki, S. & Vuokko, R. 1977: A dynamic approach to the analysis of daily height growth of plants. — Oikos 28:234–241.

Hari, P., Leikola, M. & Räsänen, P.K. 1970: A dynamic model of daily height increment of plants. — Ann. Bot. Fennici 7:375–378.

Hari, P. & Leikola, M. 1974: Further development of dynamic growth model of plant height growth. — Flora (Jena) 168:370–375.

Havas, P. 1971: The water economy of the bilberry (Vaccinium myrtillus) under winter conditions. — Rep. Kevo Subarctic Res. Stat. 8:41–52.

Havas, P. 1977: Kasviekologian perusteet. — Oulun yliopiston kasvitieteen laitoksen monisteita 1:1–153. Oulu.

Havas, P. & Kubin, E. 1983: Structure, growth and organic matter content in the vegetation cover of an old spruce forest in Northern Finland. — Ann. Bot. Fennici 20: 115–149.

Heino, R. & Hellsten, E. 1983: Tilastoja Suomen ilmastosta 1961–80. — Liite Suomen meteorologiseen vuosikirjaan, nide 80, osa 1a. 560 pp. Ilmatieteen laitos.

Ilmatieteen laitos. Kuukausikatsaus Suomen ilmastoon. Kesäkuu 1986.

Kellomäki, S. 1977: Dynamics of dry matter production in forest ground cover communities with special reference to their successional development. — Helsingin yliopiston metsänhoitotieteen laitos. Tiedonantoja 16:1–50.

Koski, V. & Selkäinaho, J. 1982: Experiments on the joint effect of heat sum and photoperiod on seedling of Betula pendula. — Commun. Inst. Forestalis Fenniae 105:1–34.

Lanner, R.M. 1964: Temperature and the diurnal rhythm of height growth in pines. — J. Forestry 62:493–495.

Moisejeva, V.Z. & Moisejev, R.G. 1977: Certain characteristic features of Vaccinium myrtillus L. formations of the northern slopes of the Ukrainian Carpatian. (In Russian). — Rastitelnye resursy 13:99–103.

Nienstaedt, H. 1974: Genetic variation in some phenological characteristics of forest trees. — In: Lieth, H. (ed.), Phenology and seasonality modeling: 389–400. Springer-Verlag, N.Y.

Purasmaa, A. 1969: Mustikan ja puolukan muuntelusta Inarin ja Utsjoen seudulla erikoisesti lehtianatomian perusteella. — Cand. phil. thesis (in the library of the Dept of Botany, Univ. of Helsinki).

Raatikainen, M., Rossi, E., Huovinen, J., Koskela, M.-L., Niemelä, M. & Raatikainen, T. 1984: Metsä- ja suomarjasadot Väli-Suomessa. — Silva Fennica 18:199–219.

Reuter, M. 1951: Einige Beobachtungen und Vergleiche betreffs des Einflusses des Breitengradens auf den Zeitpunkt des Bluhens und der Fruchtreife bei den Pflanzen. — Soc. Sci. Fennica Comment. Biol. 13(2): 1–23.

Reuter, M. 1957: Pflanzenökologische Beobachtungen in Finnland 1951–55. — Bidrag till kännedom af Finlands natur och folk. 100(1): 1–63.

Sarvas, R. 1974: Investigations on the annual cycle of development of forest trees. II. Autumn dormancy and winter dormancy. — Commun. Inst. Forestalis Fenniae 84(1): 1–101.

Snigirev, G.S. & Raptunovich, E.S. 1980: The phenology and crop formation of bilberry, cranberry and lingonberry in broadleaf–pine forest subzone in Belorussia. (In Russian)— Rastitelnye resursy 16 I: 329–334.

Solantie, R. 1981: Mustikkasadon riippuvuus kukkimisajan halloista ja kesän sademääristä kesän 1980 havaintojen valossa Suomessa. — Maataloushallinnon aikakauskirja 3:26–31.

Solantie, R. 1983: Suomen luonnonmarjat, ilmasto ja sää. — Metsäntutkimuslaitoksen tiedonantoja 90:38–49.

Stern, K. & Roche, L. 1974: Genetics of forest ecosystems. — 330 pp. Springer Verlag, Berlin.

Stewart, W.S. & Bannister, P. 1973: Seasonal changes in carbohydrate content of three Vaccinium spp. with particular reference to V. uliginosum and its distribution in the British Isles. — Flora (Jena) 162:134–155.

Tuhkanen, S. 1980: Climatic parametres and indices in plant geography. — Acta Phytogeogr. Suecica 67:1-105.

Turesson, G. 1930: The selective effect of climate upon the plant species. — Hereditas 14:99–152.

Walter, J. & Soos, J. 1963: Shoot growth patterns of some British Columbia conifers. — Forest Sci. 9:73–85.

Wilhelmi, Th. 1957: Rhythmus und Potenz des Höhenwachstum junger Nadel- und Laubhölzer. — Allg. Forstzeitschr. 12:41–47.

Woodward, F.I. 1986: Ecophysiological studies on the shrub Vaccinium myrtillus L. taken from a wide altitudinal range. — Oecologia 70:580–586.

Received 20.VII.1987

Acta Bot. Fennica 136:61–63, 1988

# Yield of the red raspberry, Rubus idaeus, after prescribed burning at Evo, southern Finland

HARRI VASANDER

Vasander, H. 1988: Yield of the red raspberry, Rubus idaeus, after prescribed burning at Evo, southern Finland. — In: Vänninen, I. & Raatikainen, M. (eds.), Proceedings of the Finnish-Soviet symposium on timber forest resources in Jyväskylä, Finland, 25–29 August 1986: 61–63. Acta Bot. Fennica 136. Helsinki. ISSN 0001-5369 ISBN 951-9469-30-3

In 1983 the yield and biomass of the red raspberry, *Rubus idaeus* L., were measured one to nine years after prescribed burning at Evo, Lammi, southern Finland. The red raspberry was most abundant two and three years after prescribed burning. One year after burning the occurrence of shoots was patchy, and in later years shoots usually occurred beside stones and stumps. Flowering was greatest two years after burning. Berry production was low. The highest values (11 kg/ha) were measured two years after burning in an area which had stones and a fairly thick humus layer. There was also great variation between areas burned in the same year. The low berry yields were due to the normal silvicultural practice of treating the grass and shrubs with herbicides.

Key words: Red raspberry, *Rubus idaeus*, production, prescribed burning

*Harri Vasander, Department of Peatland Forestry, University of Helsinki, Unioninkatu 40B, SF-00170 Helsinki, Finland*

## INTRODUCTION

In the forest regeneration chain, prescribed burning could be compared to light tillage. The aim of burning the slash, ground layer and thin surface humus layer is to prepare the site for the next tree generation (cf. Vasander & Lindholm 1985). The use of prescribed burning in forestry in Finland was most widespread in the middle of the 1950s, when approximately 30 000 ha were burned annually. There was a drastic decrease in the burned area in the 1960s (500–1 000 ha/a), but nowadays it is increasing. During summer 1986, for example, the National Board of Forestry conducted prescribed burning on an area of 2 500 ha in state forests (K. Pelkonen, pers. comm. 1987).

In the secondary succession from burned areas to mesic forest sites in southern Finland, the red raspberry, *Rubus idaeus* L., belongs to the same early developmental stage as e.g. *Epilobium angustifolium* L. and *Calamagrostis arundinacea* (L.) Roth. The more important cowberry, *Vaccinium vitis-idaea* L., and blueberry, *V. myrtillus*, become abundant later, the cowberry earlier than the blueberry (e.g. Hertz 1932, Sarvas 1937, Sirén 1955, Kardell 1980, Bocharov et al. 1986, Lindholm & Vasander 1987).

This study is part of the study project "Nutrient ecology and structure of plant communities of the *Myrtillus* type forest after prescribed burning" conducted during 1982–1984 at Evo, Lammi, southern Finland (Ruuhijärvi et al. 1983, 1986). Besides the development of vegetation on burned areas, the berry yields of *Rubus idaeus* were studied in 1983.

## MATERIALS AND METHODS

The study area belongs to the southern boreal vegetation zone (for details of the area, see Vasander & Lindholm 1985, Lindholm & Vasander 1987). Nine areas, burned one to nine years earlier, were included in the study (Table 1).

The numbers of red raspberry shoots, buds, green fruits, berries and empty receptacles were counted on three round (r= 10 m) sample plots on each area. All the berries were picked and weighed in order to calculate their mean weight. This was multiplied by the total number of buds, green berries, berries and receptacles. During the flowering time in June, a shoot analysis was made on some of the same areas and the number of flowers and buds on the shoots was counted. The abundance of *Rubus idaeus* was also estimated in vegetation and biomass analyses of the areas (Lindholm & Vasander 1987).

## RESULTS

The number of shoots varied between 2 000 and 147 600 per ha. Their biomass was less than 10 g/m$^2$. The flowering percentage decreased from two years after burning and the berry yield was greatest (11 kg/ha) two years after burning. Generally the yields were small (0–2.2 kg/ha, Table 1).

Table 1. *The study areas in Evo, southern Finland and the number of shoots, their biomass, flowering percentage and berry yields on them.* n.d.=not determined.

| Site | Felling | Burning | Years since burning | Other treatments | Shoots/ha | Biom. ($g/m^2$) | Flower % | Berry yield (kg/ha) |
|---|---|---|---|---|---|---|---|---|
| 1. Perikulma | 1981 | 1982 | 1 | | 2 000 | n.d | 0 | 0 |
| 2. Iso Keltajärvi | 1980 | 1982 | 1 | | 31 000 | 1.2 | n.d. | 0.8 |
| 3. Iso Keltajärvi | 1980 | 1981 | 2 | | 147 600 | 8.8 | 93 | 11.0 |
| 4. Latomäenmaa | 1980 | 1981 | 2 | Cardoprim 1983 | 11 600 | n.d. | n.d. | 0.1 |
| 5. Myllykorvenmaa | 1979 | 1980 | 3 | Round-up 1981 | 23 500 | 0.1 | 85 | 2.2 |
| 6. Härkämäki | 1978 | 1979 | 4 | Cardoprim 1981 | 5 500 | n.d. | n.d. | 0.7 |
| 7. Vähä Keltajärvi | 1976 | 1977 | 6 | Cardoprim 1981 | 2 800 | n.d. | n.d. | 0.7 |
| 8. Hankajärvi | 1976 | 1977 | 6 | Cardoprim 1980 | 2 400 | 0.1 | 78 | 0.1 |
| 9. Rajajärvi | 1973 | 1974 | 9 | Round-up 1976 | 23 900 | 2.1 | 73 | 1.8 |

## DISCUSSION

The first year shoots of *Rubus idaeus* are usually sterile, flowering shoots being very rare (Vaarama 1965). The species thus first allocates resources to vegetative growth, colonizing empty areas. As shading increases, more resources are allocated to the generative organs, i.e. seeds and berries (e.g. Abrahamsson 1975, Whitney 1982).

There was also a great difference between the berry yields of areas burned at the same time. For example, on one of the two-year old areas (no. 3, Table 1), there had been no herbicide treatment, the ground was gently undulating and there were stones, beside which in moist conditions (cf. Mackerron 1982) the red raspberry could flower and grow. The other area (no. 4) was a high dry area, where the increase of the species was probably hindered by the dryness and chemical treatment.

The self-thinning phase in the red raspberry (Hudson 1959) has been noted to start three years after clear-felling (Whitney 1982). In Sweden the berry yield has been reported to be greatest three to eight years after clear-felling. The total berry yield varied from less than 100 kg/ha to 1 000–3 000 kg/ha. The maximum yield (4 600 kg/ha) was measured five years after felling on an area dominated by the red raspberry. This value was comparable to the normal yields on red raspberry plantations (7 000–10 000 kg/ha) (Ingelög 1974). The total yield of *Rubus idaeus* in Swedish forests has been estimated to be 15–20 mill. kg annually, which corresponds to 0.7–0.9 kg/ha for the whole forest area. In central Finland the mean annual berry yield in forests has been estimated at approximately 0.2 kg/ha (Raatikainen et al. 1984).

More than two years after burning at Evo, the berry yields were mostly smaller than the mean values above, which was due to the scarceness of shoots. The weight of individual shoots was also lower than one year after burning (unpubl. results of the project). The treatment of grass and bushes with herbicide (Cardoprim, Round-up) had a drastic effect on the red raspberry. It seemed to suffer more and longer than, for example, the cowberry, whose berry yield has been noted to recover some years after chemical treatment (e.g. Raatikainen & Raatikainen 1982, Etholén 1983). On some areas at Evo the cowberry was fairly abundant, while the red raspberry was absent.

It can be concluded that the burned areas at Evo were not good sites for red raspberries on account of the herbicide treatments. After chemical treatment, the vegetation of these areas showed similarities to that of drier and less fertile sites (Lindholm & Vasander 1987). There were no untreated control areas, but it is known that the red raspberry is generally abundant on burned areas (e.g. Uggla 1958). During the field work I noted that much greater berry yields could be obtained, for example, beside the forest roads leading to the burned study areas. The road banks had not been treated chemically and beside the stones were suitable moist and shady microsites for the red raspberry.

## REFERENCES

Abrahamsson, W.A. 1975: Reproductive strategies in dewberries. — Ecology 56:721–726.

Bocharov, I.V., Pankov, V.B. & Kitaigorodskii, V.E. 1986: Zavisimost' fitomassy brusniki ot proektivnogo pokrytiya. — Lesnoe hozyaistvo 10(1986): 21–22.

Etholén, K. 1983: Metsänhoidollisten toimenpiteiden vaikutuksista marjasatoihin. — Metsäntutkimuslaitoksen tiedonantoja 90:153–160.

Hertz, M. 1932: Tutkimuksia aluskasvillisuuden merkityksestä kuusen uudistumiselle Etelä-Suomen kangasmailla. (Referat: Über die Bedeutung der Untervegetation für die Verjungung der Fichte auf den südfinnischen Heideböden). — Comm. Inst. Forest Fenniae 17(4):1–206.

Hudson, J.P. 1959: Effects of environment on Rubus idaeus L. I. Morphology and development of the raspberry plant. — J. Hort. Sci. 34:163–169.

Ingelög, T. 1974: Vegetationsförändringar efter förnyelseingrepp. (Summary: Vegetation changes after regeneration measures). — Sveriges Skogsvårdsförbunds Tidskr. 72: 91–103.

Kardell, L. 1980: Occurrence and production of bilberry, lingonberry and raspberry in Sweden's forests. — Forest Ecol. Manage. 2:285–298.

Lindholm, T. & Vasander, H. 1987: Vegetation and stand development of mesic forest after prescribed burning. — Silva Fennica 21:259–278.

Mackerron, D.K.L. 1982: Growth and water use in the red raspberry (Rubus idaeus L.) I. Growth and yield under different levels of soil moisture stress. — J. Hortic. Sci. 57:295–306.

Raatikainen, M. & Raatikainen, T. 1982: Lehvästöruiskutusaineiden vaikutukset marjasatoihin. — Suomen maataloustieteellisen seuran tiedote 2:80–85.

Raatikainen, M., Rossi, E., Huovinen, J., Koskela, M.-L., Niemelä, M. & Raatikainen, T. 1984: Metsä- ja suomarjasadot Väli-Suomessa. (Summary: The yields of the edible wild berries in Central Finland). — Silva Fennica 18:199–219.

Ruuhijärvi, R., Lindholm, T. & Vasander, H. 1983: Nutrient ecology and structure of plant communities of the Myrtillus type forest after prescribed burning. — Lammi Notes 10:10–11.

Ruuhijärvi, R., Lindholm, T. & Vasander, H. 1986: Some consequences of using prescribed burning in forestry. — Lammi Notes 13:1–5.

Sarvas, R. 1937: Havaintoja kasvillisuuden kehityksestä Pohjois-Suomen kuloaloilla. (Referat: Beobachtungen über die Entwicklung der Vegetation auf den Waldbrandflächen Nord-Finnlands). — Silva Fennica 44:1–64.

Sirén, G. 1955: The development of spruce forest on raw humus sites in northern Finland and its ecology. — Acta Forest. Fennica 62:1–408.

Uggla, E. 1958: Ecological effects of fire on north Swedish forest. — Almqvist & Wiksell. 18 pp. Uppsala.

Vaarama, A. 1965: Rubus idaeus L. — Vadelma, vattu. — In: Jalas, J. (ed.), Suuri kasvikirja II: 760–763. Otava. Keuruu.

Vasander, H. & Lindholm, T. 1985: Fire intensities and surface temperatures during prescribed burning. — Silva Fennica 19:1–15.

Whitney, G.G. 1982: The productivity and carbohydrate economy of a developing stand of Rubus idaeus. — Can. J. Bot. 60:2697–2703.

Received 20.VII.1987

Acta Bot. Fennica 136:65–68, 1988

# The cranberry yield in the USSR

A. Ph. CHERKASOV

Cherkasov, A. Ph. 1988: The cranberry yield in the USSR. — In: Vänninen, I. & Raatikainen, M. (eds.), Proceedings of the Finnish-Soviet symposium on timber forest resources in Jyväskylä, Finland, 25–29 August 1986: 65–68. Acta Bot. Fennica 136. Helsinki. ISSN 0001-5369 ISBN 951-9469-30-3

Two species of cranberry grow in the USSR: *Oxycoccus* (*Vaccinium*) *palustris* Pers. and *Oxycoccus microcarpus* Turcz. ex Rupr. The latter gives a low yield (about 90 kg/ha) and is of no economic value. In dense stands, the *Oxycoccus palustris* yield reaches 1 000 kg/ha and in especially favourable conditions, in comparatively small areas, 2 600 kg/ha. In normal stands, the maximum yield of *Oxycoccus palustris* is about 500 kg/ha, and the average about 200 kg/ha. The yield is highest (200–250 kg/ha on average) in the north-western part of the USSR. Towards the east and south the yield of *Oxycoccus palustris* decreases, but in the Primorye Region of the Far East, the yield again increases. A scale based on long-term studies is presented for estimating the production of *Oxycoccus palustris*.

Key words: Berry yield, *Oxycoccus palustris*, USSR

*A. Ph. Cherkasov, All Union Research Institute of Silviculture and Mechanization of Forestry, Kostroma Research Station, Prospekt Mira 134, SU-156013 Kostroma, USSR*

## CRANBERRY YIELD IN DIFFERENT PARTS OF THE USSR

Two species of cranberry grow in the USSR: *Oxycoccus* (*Vaccinium*) *palustris* Pers. and *O. microcarpus* Turcz. ex Rupr.

The yield of *Oxycoccus palustris* within the USSR ranges from a few kilogrammes to 1 000 kg/ha. Highly productive bushes of *O. palustris* are found in south Lithuania, where the harvest is often over 1 500 kg/ha. The highest yields are mainly limited to narrow strips around lakes. In most of Lithuania, including dense cranberry stands, the maximum harvest is over 1 000 kg/ha (Cherkasov et al. 1981).

Unique yield of the cranberry were discovered in the Olonez region of Karelia by the researchers of the Karelian Biology Institute, a branch of the Academy of the Sciences of the USSR. In some plant communities in the meso-oligotrophic bog of Segezh the yield of *O. palustris* reached 1 836 and 2 622 kg/ha, in plots of 100 ha and 10 ha, respectively. Such yields are determined by both large numbers of berries per unit area and their high average weight. The potentials of *O. palustris* are even higher than this; in some microplots on hummocks in the low ridgepool complex in the Darwin Reserve, by the boundary of the flooding zone of the Rybinsk reservoir, we harvested about 1 100 berries per square metre in 1973, i.e. 5 000 kg/ha (in the test plot with some microelevations the production was 623 kg/ha in terms of berry-productive area). In productive years micro-elevations of this kind can surely be found in other regions as well, but stands producing over 1 000 kg/ha are rather rare, so that the figure mentioned above may be considered the maximum for the USSR.

Natural cranberry stands that are dense and productive are seldom even 1/2 ha in extent. Usually, the bushes are interspersed with pools, small lakes, big table-like hummocks, etc. Even when dense, the stands may be only partly productive and the real harvest of *O. palustris* is thus much less than that of the optimal stands, rarely exceeding 500 kg/ha. Productivity of this kind occurs in the south of Karelia, in the south-west of the Arkhangelsk region, in the western part of the Vologda region and in the north of the Yaroslavl region, in the Kalinin, Novgorod, Leningrad and Pskov regions, the north-western part of the Smolensk region, the north of Byelorussia and in the Baltic Republics (zone 1 in Fig. 1). According to Botch and Mazing's classification (1979), the bogs of these regions belong to the domed patterned bogs of South Karelia, the Baltic coastline zone, the Eastern-Baltic zone and the north-western part of Eastern-Europe. Recently, highly productive cranberry tracts close to the above-mentioned region stretched as far as the Ukrainian and Byelorussian woodlands. But the extensive reclamation of the bogs has considerably reduced the productive areas and the harvest of *Oxycoccus*, though there are still some tracts with a fairly high yield.

Communities consisting of low bushes, cotton grass (*Eriophorum* spp.) and white bog moss (*Sphagnum* spp.) as well as simple (forested and unforested) and complex commu-

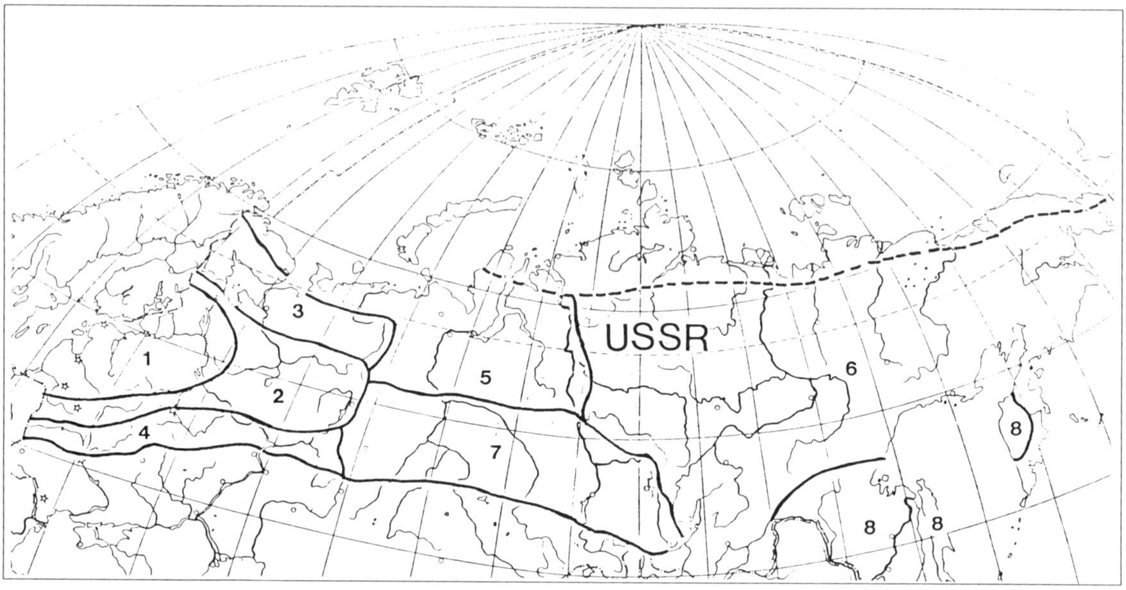


*Fig. 1.* The cranberry yield in different regions of the USSR. For explanations, see the text.

nities are very common in the above-mentioned zone. Here the average yield of *Oxycoccus* is about 200–250 kg/ha, reaching 450–500 kg/ha in dense stands.

Among the oligotrophic bogs there are sometimes meso-oligotrophic and mesotrophic communities with a higher yield: the harvest in ordinary stands averaging about 250–280 kg/ha and in dense stands 500–600 kg/ha.

To the east of this area, the yield of *Oxycoccus* decreases, averaging 180–200 kg/ha in ordinary stands and 400 kg/ha in dense ones. The zone includes the southern parts of the Arkhangelsk region and Komi ASSR (from latitude 64), the eastern part of the Vologda regions, the Perm and Kirov regions, the northern part of the Udmurt ASSR, the Mari ASSR, the northern part of the Gorki region, the Kostroma and Yaroslavl regions, the south-eastern part of the Kalinin and the north-western part of the Moscow regions, and the southern and eastern parts of the Smolensk region (zone 2 in Fig. 1). Practically all this area is part of the North-Eastern European province of domed patterned bogs and partly belongs to the zone of pine-sphagnum high and low grass bogs (Botch & Mazing 1979). According to the forest growing conditions, the area covers the subzones of the middle and southern taiga, the northern subzone of mixed forests and the northern part of the southern subzone of mixed forests.

In the northern parts of the Komi ASSR, the Arkhangelsk region, Karelia and especially the Kola Peninsula, i.e. in the northern subzone of the taiga (zone 3 in Fig. 1), the yield of *Oxycoccus* is still lower, averaging about 120–140 kg/ha (Cherkasov et al. 1981), with a maximum of 250–300 kg/ha. The relatively low yield of *O. palustris* in this area can be explained not only by the severer climatic conditions but also by considerable impurity and sometimes by the small size of the berries (the Kola Peninsula) and by the low density of the bushes, which is characteristic of the widespread pals and aapa bogs.

Near the southern boundary of the European area of the USSR (the Bryansk, Kaluga, Tula and Ryazan regions, the Mordva, Tchuvash, Tatar, Bashkir ASSR; zone 4 in Fig. 1) the *Oxycoccus* yield is also low, due to the competition of the grass vegetation characteristic of this area in rich fens and fen-like bogs (the northern monodominant forest subzone of the hardwood zone). In the oligotrophic communities with dense cranberry bushes sometimes found in this area, the *Oxycoccus* yield reaches 300 or even 500 kg/ha, but the average is under 100–120 kg.

In Siberia the *Oxycoccus* production in most of the regions is less than the European estimate in the north-western part. In the vast water-logged area stretching from the Urals as far as the Yenissey, to the north of the 58th parallel (zone 5 in Fig. 1) (the western Siberian province of the zone of domed string bog, according to Botch & Mazing 1979; the northern and middle subzones of the taiga) the average production ranges from 50 to 143 kg/ha. The relatively poor yield in this area can be explained by the unfavourable climatic conditions and low density of cranberry bushes in these bogs in which small lakes and pools sometimes cover 75% of the total area.

The Krasnoyarsk region (to the north of the Angara) is rich in palsa bogs which are not favourable for *O. palustris*. The yield here (zone 6 in Fig. 1) varies from 127 to 180 kg/ha, and is less in the Yakutsk ASSR, where plants with small-sized berries predominate.

Approximately between the 56th and 58th parallels in the area between the Urals and the Yenissey (southern taiga and northern subzone of hardwoods; zone 7 in Fig. 1) eutrophic raised bogs, fen-like bogs and rich fens, all characterized by sphagnum-moss and pine, with relatively favourable conditions for *O. palustris* are widespread. In the Omsk and Tomsk regions the *Oxycoccus* yield reaches 480 kg/ha averaging 150 kg in fen-like bogs, 100 kg in pine–sphagnum bogs and 50 kg/ha in ridge-pool complexes (Cherkasov et al. 1981).

In the Novosibirsk region (hardwood zone) and in the Krasnoyarsk region near the River Angara (southern taiga), the *O. palustris* yield varies from 13 to 309 kg/ha, the average yield in communities rich in *Oxycoccus* being 150–200 kg/ha (zone 7 in Fig. 1). As regards the area mentioned above the average yield of *O. palustris* is hardly over 100–130 kg/ha. *O. microcarpus* extends eastwards from the Urals. In the Tomsk region its yield varies from 35–53 kg/ha in eutrophic pine–sphagnum bogs to 89 kg/ha in mesotrophic bogs. This species of *Oxycoccus* predominated in the Irkutsk, Buryat, Tchita and Amur regions. Its yield is rather low, on average under 50 kg/ha. The Khabarovsk and the Primorye, the Kamchatka and Sakhalin regions (the Primorye boggy provinces of the Far East; zone 8 in Fig. 1) are known for their *O. palustris* yields, which in the Khabarovsk region in single dense cranberry communities reach 800 kg/ha, the average yield in less dense stands being 180 kg/ha. In Kamchatka the maximum yield of *Oxycoccus* is rather low, 400 kg/ha, but the average harvest is fairly high, 120–180 kg/ha (Cherkasov et al. 1981).

The yield of *O. palustris* in the European part of the USSR is thus highest in the middle and southern subzones of the taiga and in the northern and southern subzone of the hardwood forest zone.

In Siberia the yield of *O. palustris* is fairly high in the southern taiga and northern subzone of hardwood forests.

In the Far East (the Amur River territory of the Khabarovsk and of the Primorye), the conditions are favourable for *O. palustris* in the subzone characteristic of the European part of the country. The yield of *Oxycoccus* in the northern subzone of the taiga is much smaller than in the middle taiga, especially in Siberia. In the forest tundra and the tundra, the *Oxycoccus* yield is very low; moreover in boggy phytocoenoses plants with small-sized berries predominate.

## A SCALE FOR ESTIMATING THE CRANBERRY YIELD

On the basis of long-term studies and research reports made in 1972, a scale has been prepared for estimating the cranberry yield in average stands in the forest zone of the European part of the USSR (Cherkasov 1972).

The scale includes the average estimates characteristic of the highly productive north-western region and those for regions with a poor yield. In estimating the berry production of *Oxycoccus*, one must take into account that the scale is designed for normally developed phytocoenoses and should not be applied to relatively large (over 50 m$^2$) non-productive plots (bogs with pools, small lakes, etc.).

Plots less than 50 m$^2$ not covered with *Oxycoccus* or with non-productive plants are difficult to detect from a survey line and are therefore considered productive. For visual estimation of *Oxycoccus* berry production the criteria worked out by Phormosov are used with some modifications:

1) Very poor yield: single berries occur on some plants (15 berries per m$^2$, the average 9–10 berries);
2) Poor yield: single berries and small groups of them occur fairly often, but many plants do not produce berries (average 22–25 berries per 1 m$^2$);
3) Average yield: sometimes there is a considerable quantity of berries, but most of the plants produce few or no berries (average 45–50 berries per 1 m$^2$);
4) High yield: nearly half of the stands have abundant berries; the yield of the other half is poor; non-productive plants are rare (average 67–75 berries per 1 m$^2$);
5) Very high yield: rich berry production all over the plot, plants with a low yield rare or absent (average 90–100 and more berries per 1 m$^2$).

In 1977 I published another scale recommended for quantitative evaluation of *Oxycoccus* production on dense productive plots (see Table 1).

The scale enables one to compare the berry production of *Oxycoccus* on sites with different phytocoenoses in different forest regions, biotic zones, etc.

In applying the scale, one must remember that sites with single berries (average 5 berries per 1 m$^2$ or about 10 berries) should not be included in densely productive plots.

The two scales allow evaluation of the natural yield and the practical exploitation. The latter is considered to account for half of the natural yield in the USSR (see Table 1). As visual evaluation of productivity is not exact, the quantitative evaluation of *Oxycoccus* berry production varies within certain limits, which are shown in the scale.

The scales are recommended for use in forest regulation, and in the activities of harvesting organisations and forest enterprises in the forest zone of the European part of the USSR.

Table 1. *Scale for evaluation of Oxycoccus berry production in normal and dense stands in the forest zone of the European part of the USSR*. The figures above the lines denote the lower and upper limits of the natural yield of the cranberry, those under the lines the limits of exploitable yield. For more explanations, see the text.

| Stands | Relative berry production in the 1–5 scale, Yield, kg/ha: 1 | 2 | 3 | 4 | 5 |
|---|---|---|---|---|---|
| Normal | up to 70 | 71–150 | 151–250 | 251–350 | 351–500 |
| | up to 35 | 36–75 | 76–125 | 126–175 | 176–250 |
| Densely productive | up to 215 | 216–365 | 366–550 | 551–775 | 776–1000 |
| | up to 107 | 108–182 | 183–275 | 276–387 | 388–500 |

REFERENCES

Botch, M.S. & Mazing, V.V. 1979: Ecological system of bogs in the USSR. — 188 pp. Leningrad.

Cherkasov, A.Ph. 1972: Metody i sposoby prognozirovaniya srokov nastupleniya fenofaz i urozhaya dikorastushchikh yagod. — In: Produktivnost' dikorastushchikh yagodnikov i ikh khozyaistvennoe ispol'zovanie: 174–178. Kirov.

Cherkasov, A.Ph. 1977: Definition of cranberry yield and store. — Information Paper, Kostroma Information Technical Science Centre 26. 4 pp.

Cherkasov, A.Ph., Butkus, V.Ph. & Gorbunov, A.B. 1981: Klyukva (Oxycoccus). — 216 pp. Moscow.

Received 28.VII.1987

Acta Bot. Fennica 136:69–72, 1988

# Rational use, reproduction and protection of cranberries in the Latvian SSR

A. RIPA

Ripa, A. 1988: Rational use, reproduction and protection of cranberries in the Latvian SSR. — In: Vänninen, I. & Raatikainen, M. (eds.), Proceedings of the Finnish-Soviet symposium on timber forest resources in Jyväskylä, Finland, 25–29 August 1986: 69–72. Acta Bot. Fennica 136. Helsinki. ISSN 0001-5369 ISBN 951-9469-30-3

Testing and investigation of the wild cranberry plants have been carried out over the territory of the Latvian SSR. The most productive cranberry plants were to be found in fens and fen-like bogs with the average weight of berries, gathered in raised bogs, being 0.369 g, in fen-like bogs 0.422 g and in fens 0.575 g. Sixty-two botanical cranberry reserves have been established, which cover a total area of 38 600 ha. Introduction of large-fruited cranberries,*Oxycoccus macrocarpus* (Ait.) Pers., in the Latvian SSR has been started in December 1972. Cultivars of the American large-fruited cranberries are characterized by high productivity and already in the first years of fruiting the harvest was 148–2 186 g/m$^2$. A crop of 100 berries, depending upon the cultivar makes 60–193 g.

Key words: Cultivation, fertilizers, *Oxycoccus macrocarpus, O. palustris*, rational use

*A. Ripa, Botanical Garden of the Academy of Sciences of the Latvian SSR, Salaspilis, Latvian SSR, USSR*

## HABITAT CHARACTERISTICS AFFECTING CRANBERRY PRODUCTIVITY

The area of the natural stands of the cranberry *Oxycoccus palustris* Pers. (*O. quadripetalus* Br.-Bl.) has been gradually diminishing due to increasing agricultural and economic developments (extraction of peat, drainage, etc.). The main means of preserving and increasing the cranberry reserves is protection of the wild cranberry plants.

The research on wild cranberry plants carried out in the Latvian SSR has been focused on the type of wetland communities, depth of peat layer and height of water table, character of the surface of wetland (proportion of tussocks in % and their height) and the species composition of the plant cover. The cranberry crop was estimated according to a 5-point scale, ranging from 1 = scanty crop (< 90 kg/ha, with an anverage of 5 kg/ha) to 5 = excellent crop (500–700 kg/ha, with an average of 600 kg/ha). The projective cover of cranberry shoots was estimated visually according to a 10-point scale, ranging from 1-cover = < 10% to 10-cover = 90–100%.

The quantity of berries, their total and average weight and the dimensions of the largest berries were determined on test plots of 1 m$^2$. The boundaries of the cranberry stands were determined from aerial photographs. Maps of all the wetland communities were drawn in the scale of 1:100 000 and 1: 15 000, and supplemented with data obtained in the field work.

Investigations of the dependence of the depth of the peat layer upon the type of wetland showed that it was shallower in fens and fen-like bogs than in raised bogs, but that it had no influence upon the productivity of the cranberry plants. The depth of the water table in raised and fen-like bogs was 0–70 cm, in fens 0–40 cm. The average depth of the water table in raised bogs was 25 cm, in fen-like bogs 17 cm and in fens 13 cm; in the places where the productivity of the cranberry plants was highest it was 0–40 cm. The stands of the productive cranberry plants were fairly uniform. The number of tussocks in their habitats was fairly small and the tussock height average 18 cm in fens, 24 cm in fen-like bogs and 33 cm in raised bogs. The majority of tussocks were 20–30 cm high. In fens the cranberries were growing on both the tussocks and the level ground while in raised bogs they grew mainly on the tussocks.

The best cranberry plants were found in open places with no trees or in places with some pines and birches 0.5–3 cm high. The average cover of cranberry shoots in raised bogs was estimated at 3.9 points, in fen-like bogs and fens at 5.6 points. In the majority of cases, the cover of cranberry shoots was 3–4 points in raised bogs, 5–7 points in fen-like bogs and 6–8 points in fens.

The most productive cranberry plants were found in fens and fen-like bogs, the average weight of the berries gathered in raised bogs being 0.369 g, in fen-like bogs 0.422 g and in fens 0.575 g. In raised bogs, a positive correlation was observed between the cover of cranberry shoots and occurrence in the plant cover of *Eriophorum* sp. and *Andromeda polifolia* L., while a negative correlation was found with *Calluna vulgaris* L. and *Ledum palustre* L. In fen-like bogs and fens the cover of cranberry shoots was positively correlated with *Menyanthes trifoliata* L. and *Comarum palustre* L. The occurrence of *Sphagnum fuscum* (Schimper) Klinggraeff had a negative correlation with the cover of cranberry shoots, while *S. magellanicum* Bridel and *S. angustifolium* C. Jensen were observed to have a positive correlation with the cranberry cover.

## THE EFFECT OF MINERAL CONTENTS OF THE GROWTH SUBSTRATE ON CRANBERRY PRODUCTIVITY

In 1970–1974 a group of specialists from the Botanical Garden of the Academy of Sciences of the Latvian SSR inspected seven fens and fen-like bogs with natural cranberry stands (Ripa 1973, 1976). Soil samples were taken from a 0–50-cm-deep profile section in order to investigate the humidity (83.7–93.7%) of the upper soil layer and the level of the ground water. The content of potassium available to plants in the upper soil layers (0–10 and 11–20 cm) was 122.3–212 and 25–91.9 mg/100 g soil dry wt, respectively. The corresponding values for phosphorus were 19.5–32.8 and 7.3–18.1 mg/100 g. The content of organic matter in the upper soil layer (0–10 cm) in sites with productive cranberry plants varied from 93.6 to 97.7%, that of total nitrogen from 0.5 to 2.2% and the ash content from 2.4 to 6.6%. Cranberries preferred acid substrates with a pH range of 2.51–5.1.

Analysis showed that the content of available potassium in the upper soil layer (0–10 cm) of the productive cranberry reserves on fens and fen-like bogs corresponded to the normal level, while the content of phosphorus in the soil of the fens was considerably higher than in the fen-like bogs. The amount of microelements in the soil was negligible. In the upper soil layers (0–10 and 11–20 cm) the values for copper were, respectively, 2.7–10.4 and 2.5–8.0 mg/kg soil dry wt, for molybdenum 0.03–0.6 and 0.22–0.3, for manganese 25–175 and 41.6–154.3, for zinc 6.4–13.2 and for boron 0.1–1.1 mg/kg soil dry wt.

Analyses of the water of fen-like bogs during the growing season showed a very small amount of potassium and a negligible amount of phosphorus and microelements (manganese, molybdenum, boron, copper) (Ripa 1975a, b).

The influence of fertilizers upon the growth of cranberry plants in raised *Sphagnum* bogs, in fens and fen-like bogs was studied in field tests. The addition of fertilizers to the soil increased the growth and thickness of the shoots. In the Olljas (fen) and Oleru (fen-like) bogs, the difference between the control and the fertilized bogs was statistically significant when the following amounts of fertilizers were used (the numbers denote the amount of nitrogen, phosphorus and potassium in kilograms per hectare): $N_{30}P_{30}K_{30}$, $N_{20}P_{20}K_{20}$, $N_{20}P_{40}K_{20}$. Ammonium sulphate was used as a nitrogen fertilizer, potassium chloride as a potassium fertilizer, phosphoric flour and superphosphate as a phosphoric fertilizer (Ripa 1975a, 1976, Treimane 1976).

In the experiments with "Nitrophoska" (a NPK-fertilizer of granular form, $N_{44}P_{44}K_{44}$) added to the soil in the Tirelu (raised bog) and Barbeles (fen-like) bogs, small doses of fertilizers ($N_{11}P_{11}K_{11}$ and $N_{44}P_{44}K_{44}$) also gave statistically significant differences from the control in the growth of the vegetative shoots. The cytological investigations revealed no notable changes in the rate of differentiation of the floral buds which might be caused by "Nitrophoska" fertilizer. The biochemical content of the juice extracted from the berries of trial plots where fertilizers were used also within the range of variation of the juice extracted from berries of different size, which is determined by genetic factors.

In field experiments begun in 1971–1973 wild cranberry plants were limed at the rate of 1 600–2 400 kg/ha in fen-like bogs, fens and raised bogs. The results showed that liming of cranberry plants did not notably affect the growth of the shoots, but it accelerated the differentiation of the floral buds by 2–3 weeks (Ripa 1975b, 1976).

## INFLUENCE OF HERBICIDES ON THE GROWTH OF CRANBERRY

During a period of three years, the herbicide simazin was applied annually at rates of 3, 6, 12 and 24 kg/ha (Ripa 1977) to the soil of wild cranberry plants. The results showed that small doses of herbicide (3 kg/ha) were not toxical to the cranberry shoots. Moreover, the treatment with 3 kg/ha of simazin showed an increase in the growth of the cranberry shoots in comparison with the control. The application of simazin at rates of 12 and 24 kg/ha inhibited the growth of the aboveground vegetative shoots. The cytological experiments showed that 12 kg/ha of simazin delayed formation of floral buds in July by 2 weeks, in comparison with the control (Jakobsone 1975). However, the use of simazin affected neither the quality of the cranberries nor the biochemical content of the cranberry juice.

## CULTIVATION OF THE CRANBERRY IN LATVIA

Experimental cultivation of cranberries in 15 fields covering an area of 7.1 ha (Klavins 1974) has been undertaken in the Jelgava timber industry enterprise, according to a project worked out by specialists from the Projective Melioration Institute of the Latvian SSR.

Soviet and foreign scientists have come to the conclusion that the large-fruited American cranberries are the most promising for cultivation, as their productivity considerably exceeds that of bog cranberries. Besides, their upright generative shoots allow mechanical harvesting, which is very important from the economic point of view.

The climatic features of the northern regions where the large-fruited American cranberries are cultivated (British Columbia, New Brunswick, Nova Scotia, Prince Edward Island) show no essential differences from those of the Latvian SSR (Table 1).

Introduction of large-fruited American cranberries (*Oxycoccus macrocarpus* (Ait.) Pers.) into the Latvian SSR was started in December 1972 when the following cultivars were received from the University of Massachusetts: 'Black Veil', 'Scarlet', 'Early Black', 'Howes', 'Beck-with', 'Stevens', 'Bergman', 'Franklin'. The cultivars 'Me Farlin', 'Wilcox' and 'Krupnoplodnaja 1' were received the same year from the Institute of Botany of the Academy of Sciences of the Lithuanian SSR. The investigations were carried out on the trial plot of the Botanical

Table 1. *Some climatic conditions of the cranberry areas of Canada (Morgenthal 1964) and the Latvian SSR.*

| District | Aver. air T°C January | Aver. air T°C July | Aver. annual T°C | Aver. annual precipitation, mm |
|---|---|---|---|---|
| British Columbia (humid zone) | –8.0 | 16.9 | 5.0 | 694 |
| New Brunswick | –10.5 | 19.0 | 4.9 | 1 062 |
| Nova Scotia | –3.5 | 17.3 | 6.8 | 1 312 |
| Prince Edward Island | –7.8 | 19.0 | 5.6 | 970 |
| Kemeri | –5.2 | 16.4 | 5.2 | 559 |
| Liepaja | –2.8 | 16.7 | 6.5 | 692 |
| Ainazi | –4.8 | 16.9 | 5.4 | 588 |
| Daugavpils | –6.6 | 17.6 | 5.4 | 648 |

Garden and on the experimental plantation area of the Jelgava timber industry enterprise.

Depending upon the cultivar, the vegetative growth (the appearance of the green tip) of the large-fruited cranberries began on 11–20 May, the flowering period began on 11–20 June, and the ripening period on 11–31 August, i.e. the development was 10–15 days delayed in comparison with the native cranberries. This confirms the conclusion that the climatic conditions in the Latvian SSR are favourable for the development of the large-fruited American cranberries.

Soil analyses showed that the sand brought to the Kalnciems cranberry plantation has a slightly acid reaction, while the peat has a highly acid reaction. The content of easily assimilable phosphorus in the sand was very small, as was also that of easily assimilable potassium, while the contents of boron and molybdenum in the sand and peat were quite sufficient. Peat has a higher content of easily assimilable potassium and a small content of easily assimilable phosphorus.

Seedlings of the large-fruited American cranberries, planted in 1977 in the Kalnciems experimental plantation, with the above-mentioned applications of fertilizers and ground water levels, showed good growth and development. The upright shoots reached a height of 7.0–17.7 cm, the projective cover was 100% and the number of upright shoots was 1.8–3.1/m$^2$. Rooted cuttings (5–7 cm long), planted at intervals of 50 cm in rows 50 cm apart, become productive in 5–6 years. The cultivars of the large-fruited American cranberries are characterized by high productivity and even the first years of fruiting the harvest is 148–2 186 g/m$^2$ depending upon the cultivar. The weight of 100 berries is 60–193 g, depending upon the

Table 2. *Germination of the large-fruited cranberries on substrates of different pH and different types of substrate, numbers of germinated seeds (4 replicates with 25 or 50 seeds in each treatment).*

| Type of substrate | pH of substrate | Cultivar | In 2 weeks x ± S.D. | In 4 weeks x ± S.D. | % |
|---|---|---|---|---|---|
| | 2.6 | | 21.7 ± 2.2 | 22.0 ± 2.0 | 88 |
| | 3.0 | | 21.7 ± 2.8 | 21.8 ± 2.8 | 87 |
| | 3.4 | | 20.8 ± 3.5 | 21.0 ± 3.1 | 84 |
| | 4.0 | | 16.0 ± 2.7 | 16.0 ± 2.7 | 64 |
| | 4.4 | | 19.3 ± 3.4 | 19.3 ± 4.3 | 77 |
| | 5.0 | | 9.8 ± 6.6 | 10.0 ± 6.7 | 40 |
| | 5.6 | | 20.5 ± 2.5 | 21.5 ± 1.2 | 86 |
| | 6.0 | | 19.0 ± 3.1 | 20.0 ± 2.9 | 80 |
| | 6.8 | | 23.0 ± 1.4 | 23.8 ± 0.4 | 95 |
| | 7.3 | | 15.8 ± 0.9 | 16.5 ± 1.2 | 66 |
| | 7.8 | | 22.0 ± 2.1 | 22.8 ± 1.7 | 91 |
| | 8.3 | | 17.8 ± 2.2 | 19.3 ± 2.1 | 77 |
| *Sphagnum* moss | | 'Wilcox' | 44.0 ± 8.4 | 44.0 ± 8.4 | 88 |
| | | 'Early Black' | 47.0 ± 1.0 | 47.0 ± 1.0 | 94 |
| Bedding peat | | 'Wilcox' | 42.5 ± 3.5 | 44.0 ± 4.2 | 88 |
| | | 'Early Black' | 45.0 ± 1.4 | 45.5 ± 2.1 | 91 |
| *Sphagnum* moss + peat (1:1) | | 'Wilcox' | 45.5 ± 2.1 | 46.0 ± 1.4 | 92 |
| | | 'Early Black' | 44.0 ± 4.2 | 44.0 ± 4.2 | 88 |
| Filter paper | | 'Wilcox' | 42.0 ± 4.2 | 42.5 ± 4.9 | 85 |
| | | 'Early Black' | 41.0 ± 4.2 | 42.5 ± 2.1 | 85 |

cultivar. The weight of 100 berries also depends on the general conditions in the growing season and the density of the berries in the bush.

The biochemical composition of the berries of the large-fruited plants is characterized by a considerable content of pectins (0.68–0.71%) and anthocyanins (150–162 mg %). However, the large-fruited American cranberries have a lower content of free acids (2.4%) than the local wild cranberries (2.9–3.8%).

The large-fruited American cranberries are easy to propagate by green cuttings and seeds (Table 2). The green cutting should be prepared during the period of most intensive growth (June–July). The length of the cutting should be 5–7 cm. They can be planted in hot beds or in the open field, with one leaf above the surface of the bedding peat, at intervals of 2–5 cm. Seeds of the large-fruited American cranberries germinated very well, after stratification of the fresh berries for 150 days at a temperature of +4°C. After 14 days, 9–23 seeds had germinated out of 25, depending upon the reaction of the medium (pH) (Table 2), and 42–46 seeds had germinated out of 50, depending upon the type of substrate. The seedlings grew well in bedding peat and within 60 days the shoots reached a length of 10.3–21.6 cm.

## PROTECTION OF THE CRANBERRY IN LATVIA

Sixty-two botanical cranberry reserves have been established, which cover a total area of 38 800 ha. Alteration of the habitat and water regime, extraction of peat and minerals, felling of trees and destruction of plants and plant communities are strictly forbidden in these reserves. Rules have been prescribed for cranberry picking in the Latvian Republic, e.g. it is not allowed before 10 September, and tools that might damage the cranberry plants are forbidden.

## REFERENCES


Jakobsone, L. 1975: Dzervenu (Oxycoccus quadripetalus Gilib.) ziedpumpuru diferencesanas fenologija uz dazada agrofona Latvijas PSR. — In: Tautsaimnieciba derigo augu agrotehnika: 50–55. Riga.

Klavins, I. 1974: Eksperimentala dzervenu plantacija. — Zinatne un tehnika 12 (1974): 3–4.

Morgenthal, J. 1964: Die Nadelgehölze: 311–312. — Jena.

Ripa, A. 1973: Vides ekologiskie faktori dazos lielo dzervenu (Oxycoccus quadripetalus Gilib.) purvos Latvijas PSR. — In: Tautsaimnieciba derigo augu agrotehnika un selekcija: 19–44. Riga.

Ripa, A. 1975a: Meslojuma ietekme uz dzervenu (Oxycoccus quadripetalus Gilib.) augsanu augsto sunu, parejas un zalu purvos. — In: Tautsaimnieciba derigo augu agrotehnika: 14–35. Riga.

Ripa, A. 1975b: Augsnes kalkosanas ietekme uz dzervenu (Oxycoccus quadripetalus Gilib.) augsanu un razibu. — In: Tautsaimnieciba derigo augu agrotehnika: 36–46. Riga.

Ripa, A. 1976: Usloviya proizrastaniya i vliyanie udobrenii, izvestkovanii i gerbitsida na rost i razvitie klyukvy (Oxycoccus quadripetalus Gilib.). — Izv. AN LatvSSR 1976(7): 107–117.

Ripa, A. 1977: Vliyanye gerbitsida simazina na rost, razvitie i urozhainost' klyukvy chetyrekhlepestnoi. — In: Yankyavichus, K. (ed.), Klyukva: 53–56. Vilnjus.

Treimane, Sk. 1975: Mineralmeslu ietekme uz savvalas dzervenu (Oxycoccus quadripetalus Gilib.) augsanu. — In: Tautsaimnieciba derigo augu agrotehnika: 47–49. Riga.


Received 18.V.1987

Acta Bot. Fennica 136:73–76, 1988

# On the European cranberry in Soviet Karelia

V. F. YUDINA

Yudina, V. F. 1988: On the European cranberry in Soviet Karelia. — In: Vänninen, I. & Raatikainen, M. (eds.), Proceedings of the Finnish-Soviet symposium on timber forest resources in Jyväskylä, Finland, 25–29 August 1986: 73–76. Acta Bot. Fennica 136. Helsinki. ISSN 0001-5369 ISBN 951-9469-30-3

The paper gives the results of long-term studies on the phenological development of the cranberry (*Oxycoccus palustris* Pers.) and the yield dynamics. The classification of forms according to the fruit shape is presented. The effect of mineral fertilizers on the growth and productivity of the cranberry was investigated. Fertilizers were found to exert a significant effect on the vegetation in all the treatments; the coverage of sedges and dwarf shrubs increased, and an increase was observed in the shoot and leaf size of these plants. The cranberry yield was greater on the fertilized sites, but the abundant growth of sedges and dwarf shrubs caused by the fertilizers suppressed the less competitive cranberry. Removal of these competitors of the cranberry favours its growth and productivity. Therefore in order to increase the cranberry productivity, application of fertilizers must be combined with the removal of sedges and dwarf shrubs.

Key words: Berry yield, mineral fertilizers, *Oxycoccus palustris*, phenology

*V.F. Yudina, Institute of Biology, Academy of Sciences of the USSR, Karelian branch, Pushkinskaja 11, SU-185610 Petrozavodsk, USSR*

## MORPHOLOGY OF THE BERRIES

The European cranberry, *Oxycoccus palustris* Pers. (syn. *Vaccinium oxycoccus* L.), is a common and very popular plant species throughout Soviet Karelia. The following bogs are typical habitats of the cranberry: dwarf shrub–cotton grass–*Sphagnum* with or without sedge, mesotrophic dwarf shrub–sedge–*Sphagnum*, tree–dwarf shrub–herb–*Sphagnum*, oligotrophic ridge-pool bogs (Elina 1971).

Studies carried out on the bogs showed a wide variety of berries differing in shape, colour, etc. On the basis of the literature and his own data, Tokarev (1979) has developed a classification of cranberry forms.

Seven main groups were distinguished according to fruit shape: round, flattened, oblong, cubiform, pyriform, conoid and diamond-shaped (Fig. 1). The round form includes berries whose height differs insignificantly (1–2 mm) from their width, so that they are apple-shaped, cordate, ovate, etc. Berries whose width is 2 mm greater than their length are referred to the flattened group; when the length of the fruit is 2 mm greater than the width, the berries are included in the oblong group. In each group smaller units are distinguished, depending on minor characteristics of the berry shape.

## PHENOLOGY OF THE CRANBERRY

Besides routine investigations, long-term observations were made on the seasonal development of the cranberry and its yield on permanent plots located in different environmental condi-

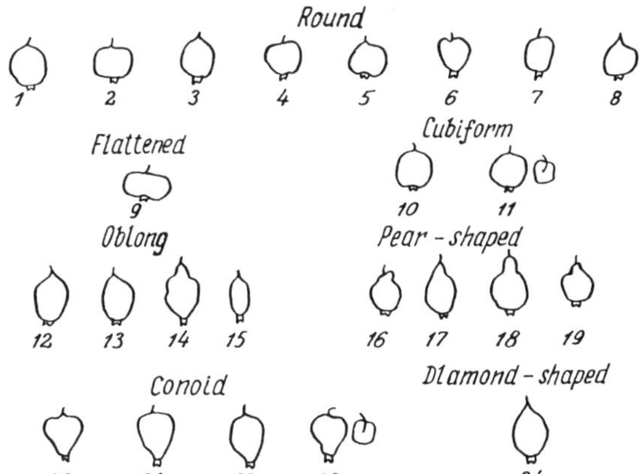


*Fig. 1:1–24.* Classification of the forms of the European cranberry (according to Tokarev 1979). — 1: round, 2: round-flattened, 3: round–oval, 4: apple-shaped with a hollow near pedicle, 5: apple-shaped, 6: apple-shaped with lengthening towards calyx, 7: egg-shaped, 8: inverted-egg-shaped, 9: turnip-shaped, 10: cubiform, 11: quadrangular, 12: oval with rough surface, 13: oval, 14: pomegranate-shaped, 15: Correlian cherry-shaped, 16: short-pear-shaped with a hollow, 17: short-pear-shaped with a sharp end, 18: pear-shaped, 19: short-pear-shaped with rough surface, 20: low conoid, 21: wedge-shaped, 22: sweet-brier-shaped, 23: pyramidal, 24: diamond-shaped.

tions: in forested dwarf shrub–herb–*Sphagnum*, sedge–*Sphagnum* and dwarf shrub–*Sphagnum* biocenoses.

The growth period of the European cranberry begins on 16 May with slight swelling of the flower-buds (Table 1). At this time most overwintering leaves have anthocyanic winter colouring. The leaves become completely green in late May; when spring was early (1977) they became green on 18 May, when it was late (1978) on 9 June. The leaf buds swell and start to unroll 8 days later than the flower-buds.

The average date of budding was recorded as 23 May, mass budding starting on 29 May. In warm years (1983) the first buds were observed much earlier, on 11 May. The latest date of mass budding was recorded in 1980 (6 June). The budding phase is of very long duration, on the average 41.9 ± 2.5 days. The first flowers generally appear on 16 June and mass flowering starts on 26 June (Table 1). The earliest flowering (31 May) was recorded in 1975, when the spring was very early; the latest flowering was observed in 1976 (25 June). The flowering period, like the budding time, is of long duration, averaging 26 ± 1.5 days. Scattered flowers are seen until the last third of July. No reflowering was observed in 1975–1983. The duration of the flowering phase is in direct relation to the weather conditions. The shortest period of flowering (17 days) was recorded in 1979, the longest (44 days) in 1978.

The first ovaries generally appear on 26–29 June, the first ripe fruits on 6–9 September. In 1975 the first ripe fruits were observed 16 days earlier than usual (20 August). In 9 study years, mass ripening of fruits was recorded on 19 September. In warmer years (1975–1980) it occurred in the first third of September, and in cooler years in late September.

In order to determine the times of berry ripening, one should know the time elapsing between the start of different phases. For example, there are 81.2 ± 2.6 days between the beginning of flowering and the appearance of the first ripe fruits. This time varies depending on the weather conditions; the shortest period (61 days) was recorded in the fairly warm summer of 1979, the longest (98 days) in 1983. The time between the beginning of flowering and mass fruiting averages 95.2 ± 2.2 days, ranging from 80 to 122 days (Yudina & Maksimova 1982). These data make it possible to forecast the approximate times of berry ripening.

Although the times of the beginning of the different phases vary with the year, they correspond to a certain sum of effective temperatures. Swelling of the flower-buds starts at a sum of 49 ± 4 day-degrees (> 5°C); mass budding begins at 132 ± 9; flowering at 254 ± 12; mass flowering at 346 ± 10; the appearance of ripe fruits at 985 ± 34; mass fruit formation at 1 037 ± 35 (Table 1).

## FACTORS AFFECTING THE BERRY YIELD OF THE CRANBERRY

The size of the cranberry yield depends on the number of flowers. These are formed in the year preceding fruiting. Therefore, the yield is primarily dependent on the weather conditions in the previous year. In favourable years much more buds are formed. By the method of dispersion analysis, Tokarev (1982) has shown that weather conditions have a more important effect on the productivity of the cranberry (strength of the effect 38–47%) than ecophytocenotic factors. The flowering period is also very significant in the life of the cranberry. As has been noted, flowering occurs in June and is of rather long duration, averaging 26.9 ± 1.5 days in the communities studied. According to our data, the duration of blooming of one flower (from unrolling of corolla to its falling off) varied from 5 to 10 days, depending on the weather conditions. In 1976 the average length of flowering of one cranberry flower was 10.8 ± 0.1 days, in 1979 — 7.6 ± 0.2, in 1980 — 4.5 ± 0.8, in 1981 — 5.3 ± 0.8 days. The minimum duration (2 days) was observed in 1979–1981, the maximum (14 days) in the cool and wet season of 1976.

The weather during flowering and formation of the ovary significantly affects the yield of berries. In 1975–1982 the ratio between ripe fruits and flowers was 32–90%. Especially high flower fall (87–90%) was observed in 1982, when the weather during flowering was characterized by frequent rain and low air temperatures. It should also be noted that almost every year a

Table 1. *Variation in the seasonal development of the European cranberry in 1975–1983.* The dates given are the earliest and the latest recorded day, and the mean date ± S.E. The sum of the effective temperatures includes the minimum, maximum and the mean ± S.E.

| Phases and subphases | Date | | Sum of effective temperatures (> +5° C) | |
|---|---|---|---|---|
| | Range | Mean ± SE | Range | Mean ± SE |
| Complete greening of overwintering leaves | 18.V. – 9.VI. | 30.V. ± 2.4 | 98 – 181 | 143 ± 12 |
| Swelling of vegetative buds | 10.V. – 31.V. | 24.V. ± 2.2 | 77 – 127 | 97 ± 6 |
| Beginning of shoot growth | 16.V. – 2.VI. | 27.V. ± 1.7 | 98 – 164 | 123 ± 9 |
| Swelling of flower buds | 3.V. – 26.V. | 16.V. ± 2.2 | 28 – 61 | 49 ± 4 |
| Beginning of budding | 11.V. – 2.VI. | 23.V. ± 2.2 | 60 – 127 | 93 ± 8 |
| Mass budding | 18.V. – 6.VI. | 29.V. ± 1.9 | 98 – 164 | 132 ± 9 |
| Beginning of flowering | 31.V. – 25.VI. | 16.VI. ± 2.3 | 205 – 315 | 254 ± 12 |
| Mass flowering | 12.VI. – 9.VII. | 12.VI. ± 2.6 | 291 – 390 | 346 ± 10 |
| End of flowering | 25.VI. – 2.VIII. | 12.VII. ± 3.1 | 456 – 543 | 503 ± 10 |
| First ripe fruits | 20.VIII. – 15.IX. | 6.IX. ± 2.1 | 846 – 1174 | 985 ± 34 |
| Mass ripening of fruits | 8.IX. – 30.IX. | 19.IX. ± 2.2 | 874 – 1202 | 1037 ± 35 |

Table 2. *Dynamics of cranberry productivity in a mesotrophic sedge–Sphagnum community in Soviet Karelia.*

| Year | Number of berries/m$^2$ | Yield kg/ha |
|---|---|---|
| 1972 | 30.4 | 233 |
| 1973 | 134.0 | 813 |
| 1974 | 105.7 | 750 |
| 1975 | 16.0 | 113 |
| 1976 | 11.5 | 52 |
| 1977 | 6.5 | 46 |
| 1978 | 62.5 | 423 |
| 1979 | 23.4 | 163 |
| 1980 | 38.8 | 208 |
| 1981 | 20.4 | 176 |
| Mean | 44.9±12.6 | 296±83 |

great number of flowerbuds are damaged by early summer frosts. In 1975, for example, 60–80% of the flower-buds were destroyed by frost and flowering was very poor on all the sites in that year.

August is characterized by an intensive increase of fruit weight. The increase of the weight occurs at the expense of the linear growth parameters of the berries (Chesnokov 1982). In 1980 frosts in late August damaged a great number of green berries; in all the study sites about 50% of the berries could not be used. This phenomenon was not observed in other years.

The yield of the cranberry thus varies significantly between years, depending on the weather and other factors (Table 2). As is seen from Table 2, in a mesotrophic sedge–*Sphagnum* community the yield varied from 46 kg/ha in a poor year to 813 kg/ha in a good year. The ecophytocenotic conditions have an effect, but it is not very pronounced. Studies on the dynamics of the yield in various ecophytocenotic conditions have shown that it is greatest in mesotrophic sedge–*Sphagnum* communities situated near swamps; the average productivity here is 296 ± 86 kg/ha. Fairly high productivity, (200 ± 58 kg/ha) is observed in oligotrophic dwarf shrub–sedge–cotton grass–*Sphagnum* communities at mire margins with occasional pine and birch (Yudina et al. 1983).

## INFLUENCE OF MINERAL FERTILIZERS ON THE GROWTH AND BERRY YIELD

The influence of mineral fertilizers on the growth and productivity of the European cranberry was studied in mesotrophic *Sphagneta magellanici + S. falaxi* bog facies in sedge–*Sphagnum* and dwarf shrub–*Sphagnum* communities. The experiment comprised five treatments,

I: the control (no fertilizers),
II: $N_{30}P_{60}$
III: $N_{30}K_{45}$
IV: $P_{60}K_{45}$
V: $N_{30}P_{60}K_{45}$

Fertilizers in the form of ammonium nitrate, double superphosphate and chloric potassium were applied annually in late May–early June in 1976–1979.

Table 3. *Yield of berries (g/m$^2$) of European cranberry in plots treated with mineral fertilizers.* For the treatments (I–V), see the text.

| Year | Treatments I | II | III | IV | V |
|---|---|---|---|---|---|
| 1976 | 19 ± 2 | 23 ± 2 | 16 ± 2 | 16 ± 2 | 19 ± 2 |
| 1977 | 29 ± 3 | 24 ± 4 | 15 ± 2 | 34 ± 4 | 24 ± 4 |
| 1978 | 19 ± 2 | 12 ± 2 | 11 ± 2 | 24 ± 3 | 28 ± 4 |
| 1979 | 6 ± 1 | 5 ± 1 | 3 ± 1 | 8 ± 1 | 9 ± 1 |
| 1980 | 11 ± 1 | 4 ± 1 | 6 ± 1 | 5 ± 1 | 9 ± 1 |
| 1981 | 38 ± 4 | 13 ± 2 | 18 ± 4 | 11 ± 3 | 8 ± 2 |
| 1982 | 6 ± 1 | 4 ± 1 | 15 ± 3 | 15 ± 3 | 9 ± 2 |

The fertilizers affected the vegetation significantly; the coverage of sedges and dwarf shrubs increased 1.5-fold and an increase in shoot and leaf size was also observed. In all the treatments the coverage of the cranberry increased from 7 to 17–25%, the number of shoots being much higher than in the control. In the years 1976–1978 the number of creeping shoots in the experimental treatments was 22–69% higher and that of upright shoots was 11–38% higher. The cessation of the application of fertilizers affected the growth of the cranberry. Thus in treatments II and III the fertilizers were last applied in 1978 and in 1979 the growth of the cranberry in these treatments was at the level of the control or lower. In treatments IV and V application of fertilizers was stopped in 1981, and this immediately affected the growth of the cranberry.

The berry yield in the fertilizer experiment is presented in Table 3. In the first year fertilizers could affect only the growth of the shoots, because the reproductive organs had formed in the previous year. The influence of fertilizers on the yield could become apparent only in 1977.

In the treatments with NP (II) and NK (III) fertilizers, the productivity was much lower than in the control (I). An increase of the yield was observed in the treatments with PK (IV) fertilizers applied in 1977–1979 and with the NPK (V) in 1978–1979. From 1980 onwards the number of berries was much lower in all the treatments than in the control. This seems to be due to the enhanced growth of sedges and dwarf shrubs, which strongly suppressed the less competitive cranberry.

The fertilizers did not increase the size or weight of the fruit; in all the study years the average weight (100 berries) of the fruit was lower in the experimental sites than in the control.

The contribution of mineral fertilizers to the cranberry nutrition was studied by determining the content of nitrogen, phosphorus and potassium in the leaves and stems.

Comparison of the leaves and stems in the different treatments showed that their element concentrations increased with the application of fertilizers. The amount of mineral elements in various organs of the cranberry depends on the fertilizers applied, this dependence being particularly clear for nitrogen and phosphorus. In the treatments in which these elements were applied, their amount in the leaves was much larger. In the case of potassium such a relation was not observed.

Table 1. *Number of basidiocarps collected in different stands.* Development class: T: sapling stands, R: pole stage stands, H: thinning stage stands and P: stands at final felling stage. Vertical lines separate groups of stands, which were situated side by side and within which comparisons were made. It should be noted that the differences between stand groups do not indicate differences in the amount of fruit bodies, as the number of investigations is different.

| Stand | *1* | *2* | *2a* | *2b* | *3* | *4* | *5* | *6* | *7* | *8* | *9* | *10* | *11* | *12* | *13* | *14* | *15* | *16* | *17* | *18* | *19* | *20* | *21* | *22* | *23* |
|---|---|---|---|---|---|---|---|---|---|---|---|---|---|---|---|---|---|---|---|---|---|---|---|---|---|
| Dominant height (m) | 11 | 22 | – | – | 6 | 21 | 4 | 3 | 1 | 9 | 17 | 7 | 18 | <1 | 9 | 4 | 15 | <1 | 1 | 7 | 16 | 8 | 8 | 1 | 22 |
| Development class | H | P | P | R | R | P | T | T | T | H | P | R | P | T | R | R | H | T | T | R | H | R | R | T | P |
| Number of surveys | 11 | 11 | 6 | 6 | 9 | 9 | 9 | 9 | 9 | 9 | 9 | 9 | 9 | 9 | 9 | 7 | 7 | 7 | 7 | 7 | 7 | 7 | 7 | 7 | 7 |
| *Albatrellus ovinus* | 0 | 0 | 0 | 0 | 0 | 0 | 0 | 0 | 0 | 0 | 0 | 0 | 0 | 0 | 0 | 0 | 0 | 0 | 0 | 5 | 7 | 0 | 0 | 0 | 0 |
| *Amanita muscaria* | 1 | 0 | 0 | 0 | 7 | 1 | 3 | 5 | 0 | 3 | 0 | 1 | 0 | 0 | 10 | 1 | 1 | 1 | 1 | 0 | 1 | 0 | 1 | 4 | 0 |
| *A. porhyria* | 3 | 0 | 0 | 0 | 0 | 0 | 1 | 0 | 0 | 1 | 0 | 0 | 1 | 0 | 0 | 1 | 0 | 0 | 0 | 0 | 0 | 0 | 0 | 0 | 0 |
| *Armillariella mellea* | 2 | 0 | 0 | 0 | 1 | 0 | 0 | 0 | 8 | 0 | 0 | 0 | 0 | 0 | 0 | 1 | 0 | 0 | 0 | 7 | 0 | 2 | 14 | 0 | 6 |
| *Boletus edulis* | 2 | 0 | 0 | 0 | 0 | 0 | 0 | 0 | 0 | 0 | 0 | 3 | 0 | 0 | 1 | 0 | 0 | 0 | 0 | 0 | 0 | 1 | 0 | 0 | 0 |
| *Cantharellula umbonata* | 9 | 19 | 0 | 0 | 0 | 9 | 3 | 0 | 0 | 43 | 11 | 11 | 1 | 0 | 18 | 20 | 3 | 0 | 0 | 15 | 21 | 0 | 2 | 2 | 1 |
| *Chalciporus piperatus* | 0 | 0 | 0 | 0 | 0 | 0 | 0 | 0 | 0 | 0 | 0 | 0 | 0 | 0 | 0 | 0 | 0 | 0 | 0 | 0 | 0 | 0 | 0 | 0 | 0 |
| *Chroogomphus rutilus* | 0 | 1 | 0 | 0 | 1 | 2 | 0 | 0 | 0 | 0 | 0 | 0 | 0 | 0 | 10 | 2 | 0 | 0 | 0 | 3 | 0 | 3 | 1 | 0 | 1 |
| *Clavaria flava* | 5 | 1 | 0 | 0 | 0 | 0 | 0 | 0 | 0 | 0 | 0 | 0 | 0 | 0 | 0 | 3 | 0 | 0 | 0 | 1 | 2 | 0 | 1 | 1 | 0 |
| *Clitocybe clavipes* | 0 | 0 | 0 | 0 | 2 | 2 | 0 | 0 | 0 | 0 | 0 | 0 | 0 | 0 | 0 | 0 | 0 | 0 | 0 | 0 | 0 | 1 | 1 | 1 | 2 |
| *C.* sp. | 1 | 0 | 0 | 0 | 31 | 0 | 16 | 0 | 29 | 1 | 7 | 8 | 21 | 79 | 21 | 0 | 0 | 7 | 4 | 6 | 1 | 24 | 14 | 5 | 2 |
| *Collybia butyracea* | 0 | 0 | 0 | 0 | 0 | 0 | 0 | 0 | 1 | 1 | 0 | 1 | 1 | 0 | 0 | 19 | 1 | 0 | 3 | 0 | 0 | 5 | 0 | 0 | 0 |
| *C. dryophila* | 0 | 0 | 0 | 0 | 2 | 1 | 0 | 0 | 10 | 3 | 0 | 2 | 0 | 1 | 2 | 0 | 0 | 23 | 6 | 0 | 2 | 10 | 0 | 4 | 0 |
| *C. maculata* | 0 | 1 | 0 | 0 | 0 | 0 | 0 | 0 | 0 | 0 | 0 | 0 | 0 | 0 | 0 | 0 | 0 | 0 | 0 | 0 | 0 | 7 | 0 | 0 | 0 |
| *C. peronata* | 0 | 0 | 0 | 0 | 0 | 4 | 14 | 0 | 0 | 0 | 0 | 0 | 0 | 0 | 8 | 5 | 0 | 0 | 0 | 4 | 0 | 0 | 0 | 1 | 0 |
| *C. putilla* | 0 | 0 | 0 | 0 | 0 | 0 | 0 | 0 | 0 | 0 | 2 | 0 | 0 | 0 | 0 | 0 | 0 | 0 | 0 | 0 | 0 | 0 | 0 | 0 | 0 |
| *Coltricia perennis* | 0 | 0 | 0 | 0 | 0 | 0 | 0 | 0 | 0 | 2 | 0 | 0 | 0 | 0 | 0 | 0 | 0 | 0 | 0 | 0 | 0 | 0 | 0 | 0 | 0 |
| *Cortinarius cinnam.* coll. | 0 | 1 | 0 | 0 | 0 | 0 | 0 | 0 | 0 | 1 | 0 | 7 | 0 | 0 | 69 | 15 | 0 | 0 | 0 | 4 | 20 | 2 | 0 | 1 | 0 |
| *C. camphoratus* | 0 | 0 | 0 | 0 | 0 | 0 | 0 | 0 | 0 | 0 | 0 | 0 | 0 | 0 | 0 | 0 | 0 | 0 | 0 | 1 | 0 | 4 | 0 | 0 | 0 |
| *C. collinitus* | 10 | 2 | 6 | 0 | 0 | 2 | 0 | 0 | 0 | 4 | 7 | 0 | 1 | 0 | 0 | 1 | 2 | 0 | 0 | 1 | 0 | 1 | 0 | 1 | 0 |
| *C. gentilis* | 5 | 2 | 0 | 4 | 0 | 2 | 0 | 0 | 0 | 1 | 0 | 0 | 0 | 0 | 3 | 0 | 0 | 0 | 0 | 0 | 1 | 0 | 0 | 0 | 0 |
| *C. mucosus* | 8 | 0 | 0 | 4 | 6 | 4 | 0 | 0 | 0 | 3 | 1 | 6 | 0 | 0 | 3 | 5 | 1 | 0 | 0 | 0 | 4 | 2 | 1 | 0 | 0 |
| *C. semisanguineus* | 12 | 0 | 0 | 0 | 3 | 0 | 0 | 0 | 0 | 5 | 1 | 41 | 0 | 0 | 24 | 9 | 0 | 0 | 0 | 10 | 0 | 3 | 0 | 0 | 0 |
| *C. vibratilis* | 128 | 0 | 0 | 0 | 1 | 0 | 0 | 0 | 0 | 3 | 0 | 0 | 0 | 0 | 1 | 0 | 0 | 0 | 0 | 0 | 0 | 0 | 0 | 0 | 0 |
| *C.* sp. | 0 | 44 | 12 | 21 | 69 | 16 | 1 | 4 | 0 | 32 | 5 | 65 | 12 | 0 | 153 | 184 | 5 | 2 | 0 | 91 | 17 | 144 | 17 | 13 | 16 |
| *Cystoderma amian.* coll. | 94 | 18 | 10 | 142 | 2 | 29 | 3 | 0 | 3 | 83 | 8 | 79 | 21 | 2 | 95 | 60 | 16 | 5 | 1 | 206 | 50 | 69 | 3 | 2 | 5 |
| *Gomphidius roseus* | 0 | 0 | 0 | 0 | 1 | 0 | 0 | 1 | 0 | 0 | 0 | 2 | 0 | 0 | 2 | 0 | 0 | 0 | 0 | 0 | 0 | 0 | 1 | 2 | 0 |
| *Galerina marginata* coll. | 0 | 0 | 0 | 0 | 0 | 0 | 0 | 0 | 0 | 0 | 5 | 0 | 0 | 0 | 0 | 0 | 0 | 0 | 0 | 0 | 0 | 0 | 0 | 0 | 0 |
| *G.* sp. | 6 | 0 | 0 | 0 | 0 | 1 | 2 | 0 | 0 | 5 | 0 | 0 | 0 | 0 | 1 | 0 | 0 | 0 | 0 | 0 | 0 | 0 | 0 | 0 | 0 |
| *Hebeloma* sp. | 3 | 0 | 0 | 0 | 0 | 3 | 0 | 0 | 2 | 0 | 0 | 0 | 5 | 0 | 0 | 0 | 0 | 0 | 0 | 0 | 0 | 0 | 0 | 0 | 0 |
| *Hygrophoropsis aurantiaca* | 2 | 24 | 0 | 0 | 1 | 0 | 9 | 0 | 9 | 6 | 3 | 0 | 6 | 2 | 20 | 0 | 0 | 0 | 2 | 0 | 1 | 0 | 2 | 4 | 0 |
| *Hygrocybe conica* | 0 | 0 | 0 | 0 | 0 | 0 | 0 | 0 | 1 | 0 | 0 | 0 | 0 | 0 | 0 | 0 | 0 | 0 | 0 | 0 | 0 | 0 | 0 | 0 | 0 |
| *Hygrophorus hypotheus* | 3 | 0 | 0 | 4 | 0 | 0 | 0 | 0 | 0 | 5 | 0 | 27 | 0 | 0 | 4 | 32 | 0 | 0 | 0 | 15 | 0 | 5 | 0 | 0 | 0 |
| *Hypholoma capnoides* | 84 | 0 | 0 | 2 | 6 | 2 | 12 | 0 | 2 | 34 | 143 | 0 | 247 | 112 | 25 | 24 | 0 | 3 | 121 | 64 | 154 | 71 | 7 | 55 | 0 |
| *Inocybe* sp. | 0 | 0 | 0 | 0 | 0 | 0 | 1 | 0 | 0 | 0 | 0 | 4 | 0 | 0 | 0 | 0 | 0 | 0 | 0 | 0 | 0 | 0 | 1 | 1 | 0 |
| *Laccaria laccata* coll. | 3 | 0 | 0 | 4 | 8 | 0 | 268 | 5 | 70 | 0 | 0 | 4 | 0 | 96 | 0 | 0 | 0 | 94 | 172 | 1 | 0 | 0 | 1 | 19 | 0 |
| *Lactarius fuscus* | 0 | 0 | 0 | 3 | 0 | 0 | 0 | 0 | 0 | 24 | 0 | 0 | 0 | 0 | 0 | 0 | 0 | 0 | 0 | 5 | 0 | 1 | 2 | 4 | 1 |
| *L. musteus* | 0 | 0 | 0 | 0 | 0 | 0 | 0 | 0 | 0 | 0 | 0 | 5 | 0 | 0 | 2 | 0 | 0 | 0 | 0 | 0 | 0 | 1 | 0 | 0 | 0 |
| *L. rufus* | 270 | 149 | 36 | 125 | 135 | 149 | 62 | 37 | 0 | 422 | 61 | 456 | 111 | 33 | 338 | 213 | 86 | 3 | 44 | 206 | 102 | 164 | 34 | 50 | 94 |
| *L. torminosus* | 0 | 0 | 0 | 0 | 0 | 0 | 0 | 0 | 0 | 0 | 0 | 1 | 0 | 0 | 1 | 0 | 0 | 0 | 0 | 1 | 0 | 0 | 0 | 1 | 0 |

*(contnd.)*

The stands were divided into four groups in regard to the development class:

P: stands at the stage of final felling or over-aged stands, age over 70 years, the dominant trees partly over-aged, canopy density 45–75%, average height over 16 m.

H: stands at the thinning stage, average height ca. 10–15 m, age 30–50 years, canopy density 70–90%, most stands thinned at least once.

R: stands at the pole stage, average height 4–9 m, canopy density 80–95%, age 20–30 years, not thinned.

T: sapling stands, height less than 4 m, partly open, age 5–15 years.

The stands originated from both natural and artificial regeneration.

Table 1 (*contnd.*)

| Stand | *1* | *2* | *2a* | *2b* | *3* | *4* | *5* | *6* | *7* | *8* | *9* | *10* | *11* | *12* | *13* | *14* | *15* | *16* | *17* | *18* | *19* | *20* | *21* | *22* | *23* |
|---|---|---|---|---|---|---|---|---|---|---|---|---|---|---|---|---|---|---|---|---|---|---|---|---|---|
| Dominant height (m) | 11 | 22 | – | – | 6 | 21 | 4 | 3 | 1 | 9 | 17 | 7 | 18 | <1 | 9 | 4 | 15 | <1 | 1 | 7 | 16 | 8 | 8 | 1 | 22 |
| Development class | H | P | P | R | R | P | T | T | T | H | P | R | P | T | R | R | H | T | T | R | H | R | R | T | P |
| Number of surveys | 11 | 11 | 6 | 6 | 9 | 9 | 9 | 9 | 9 | 9 | 9 | 9 | 9 | 9 | 9 | 7 | 7 | 7 | 7 | 7 | 7 | 7 | 7 | 7 | 7 |
| *L. vietus* | 0 | 0 | 0 | 0 | 0 | 0 | 0 | 0 | 0 | 0 | 0 | 0 | 0 | 0 | 0 | 0 | 0 | 1 | 0 | 6 | 0 | 0 | 0 | 0 | 0 |
| *Leccinum scabrum* | 0 | 0 | 0 | 0 | 1 | 0 | 1 | 0 | 0 | 0 | 0 | 0 | 0 | 0 | 0 | 0 | 5 | 4 | 7 | 0 | 2 | 0 | 0 | 2 | 1 |
| *L. testaceoscabrum* | 0 | 1 | 0 | 1 | 0 | 0 | 0 | 0 | 0 | 0 | 0 | 1 | 0 | 0 | 0 | 0 | 0 | 0 | 0 | 0 | 1 | 0 | 0 | 0 | 0 |
| *Marasmius scorodonius* | 0 | 0 | 0 | 0 | 0 | 0 | 0 | 0 | 10 | 4 | 0 | 0 | 0 | 0 | 0 | 1 | 0 | 0 | 6 | 0 | 0 | 4 | 3 | 0 | 0 |
| *Mycena clavicularis* | 0 | 1 | 0 | 0 | 5 | 0 | 0 | 0 | 0 | 10 | 0 | 0 | 0 | 0 | 0 | 0 | 0 | 0 | 0 | 0 | 0 | 0 | 0 | 0 | 0 |
| *M. galericulata* | 0 | 0 | 0 | 0 | 0 | 0 | 0 | 0 | 0 | 0 | 0 | 0 | 0 | 0 | 0 | 0 | 0 | 0 | 0 | 6 | 0 | 0 | 0 | 8 | 0 |
| *M. luteoalcalina* | 3 | 0 | 0 | 0 | 1 | 0 | 0 | 0 | 0 | 1 | 0 | 0 | 0 | 0 | 0 | 0 | 0 | 0 | 0 | 0 | 0 | 0 | 0 | 0 | 0 |
| *M. rubromarginata* | 0 | 0 | 0 | 0 | 6 | 0 | 0 | 0 | 0 | 0 | 0 | 0 | 0 | 0 | 0 | 0 | 0 | 0 | 0 | 0 | 0 | 0 | 0 | 0 | 0 |
| *Paxillus atrotomentosus* | 0 | 0 | 0 | 0 | 0 | 0 | 0 | 0 | 0 | 0 | 0 | 0 | 0 | 0 | 0 | 0 | 1 | 0 | 0 | 0 | 0 | 0 | 0 | 0 | 0 |
| *P. involutus* | 2 | 0 | 0 | 0 | 0 | 0 | 0 | 0 | 0 | 3 | 0 | 1 | 0 | 7 | 0 | 1 | 0 | 2 | 3 | 0 | 0 | 0 | 1 | 0 | 1 |
| *P. panuoides* | 0 | 0 | 0 | 0 | 0 | 0 | 0 | 0 | 0 | 0 | 0 | 0 | 0 | 0 | 0 | 0 | 0 | 0 | 02 | 0 | 0 | 0 | 0 | 0 | 0 |
| *Phaeocollybia* sp. | 3 | 0 | 0 | 0 | 0 | 0 | 0 | 0 | 0 | 0 | 0 | 0 | 0 | 0 | 0 | 0 | 0 | 0 | 0 | 0 | 0 | 0 | 0 | 0 | 0 |
| *Pholiota spumosa* | 0 | 0 | 0 | 0 | 12 | 0 | 8 | 0 | 4 | 0 | 0 | 1 | 0 | 0 | 1 | 0 | 0 | 0 | 61 | 2 | 0 | 8 | 1 | 4 | 0 |
| *Pluteus* sp. | 0 | 0 | 0 | 0 | 0 | 0 | 0 | 0 | 1 | 0 | 0 | 0 | 0 | 1 | 0 | 0 | 0 | 0 | 1 | 0 | 0 | 0 | 0 | 0 | 0 |
| *Psathyrella* sp. | 0 | 0 | 0 | 0 | 0 | 0 | 0 | 0 | 0 | 1 | 0 | 0 | 0 | 0 | 0 | 0 | 0 | 0 | 0 | 0 | 0 | 0 | 0 | 0 | 0 |
| *Rhodophyllus* sp. | 17 | 3 | 9 | 6 | 0 | 20 | 3 | 0 | 0 | 24 | 16 | 18 | 6 | 0 | 19 | 20 | 3 | 0 | 0 | 34 | 51 | 6 | 0 | 0 | 8 |
| *Rozites caperatus* | 54 | 23 | 0 | 8 | 4 | 0 | 0 | 0 | 0 | 0 | 0 | 0 | 0 | 0 | 0 | 3 | 0 | 0 | 0 | 2 | 1 | 1 | 2 | 0 | 0 |
| *Russula adusta* | 2 | 0 | 0 | 0 | 0 | 3 | 1 | 0 | 0 | 0 | 1 | 0 | 2 | 0 | 7 | 0 | 0 | 1 | 4 | 1 | 0 | 4 | 2 | 2 | 4 |
| *R. aeruginea* | 0 | 0 | 0 | 0 | 0 | 0 | 0 | 0 | 0 | 0 | 1 | 0 | 0 | 0 | 0 | 0 | 0 | 0 | 0 | 1 | 0 | 0 | 0 | 0 | 0 |
| *R. decolorans* | 70 | 53 | 40 | 160 | 0 | 9 | 1 | 0 | 0 | 25 | 6 | 6 | 39 | 0 | 1 | 0 | 6 | 0 | 0 | 5 | 2 | 0 | 0 | 0 | 3 |
| *R. emetica* | 1 | 1 | 0 | 1 | 0 | 0 | 0 | 0 | 0 | 0 | 0 | 1 | 0 | 0 | 0 | 0 | 0 | 0 | 0 | 0 | 0 | 0 | 0 | 0 | 0 |
| *R. flava* | 2 | 0 | 0 | 0 | 0 | 0 | 0 | 0 | 0 | 0 | 0 | 0 | 0 | 0 | 0 | 0 | 0 | 0 | 0 | 0 | 0 | 0 | 0 | 0 | 0 |
| *R. fragilis* coll. | 9 | 4 | 0 | 0 | 3 | 0 | 0 | 0 | 0 | 11 | 0 | 0 | 0 | 0 | 0 | 2 | 0 | 0 | 0 | 0 | 0 | 0 | 1 | 0 | 2 |
| *R. paludosa* | 3 | 6 | 16 | 16 | 1 | 3 | 0 | 0 | 0 | 12 | 2 | 2 | 1 | 0 | 0 | 1 | 0 | 0 | 0 | 0 | 0 | 2 | 0 | 0 | 0 |
| *R. versicolor* | 0 | 0 | 0 | 0 | 0 | 0 | 2 | 0 | 3 | 0 | 0 | 1 | 0 | 0 | 3 | 5 | 2 | 6 | 0 | 12 | 0 | 1 | 2 | 3 | 5 |
| *R. vinosa* | 11 | 8 | 11 | 39 | 5 | 14 | 0 | 0 | 0 | 50 | 19 | 4 | 28 | 0 | 10 | 45 | 30 | 0 | 0 | 34 | 53 | 54 | 6 | 2 | 15 |
| *R. xerampelina* | 8 | 0 | 0 | 2 | 6 | 1 | 2 | 0 | 0 | 0 | 0 | 1 | 4 | 0 | 0 | 1 | 2 | 0 | 0 | 6 | 10 | 0 | 5 | 2 | 0 |
| *R.* sp. | 1 | 0 | 0 | 0 | 2 | 0 | 0 | 0 | 0 | 0 | 2 | 0 | 6 | 0 | 0 | 3 | 6 | 5 | 0 | 0 | 0 | 2 | 5 | 1 | 4 |
| *Stropharia hornemannii* | 0 | 0 | 0 | 0 | 0 | 0 | 0 | 0 | 0 | 0 | 0 | 0 | 0 | 0 | 0 | 0 | 0 | 0 | 0 | 0 | 0 | 0 | 1 | 0 | 0 |
| *Suillus bovinus* | 1 | 0 | 0 | 0 | 65 | 7 | 12 | 30 | 2 | 8 | 2 | 29 | 4 | 7 | 14 | 5 | 9 | 0 | 37 | 18 | 0 | 0 | 3 | 11 | 3 |
| *S. luteus* | 0 | 0 | 0 | 0 | 0 | 0 | 0 | 3 | 0 | 0 | 0 | 1 | 0 | 0 | 2 | 0 | 0 | 0 | 0 | 2 | 0 | 0 | 0 | 0 | 0 |
| *S. variegatus* | 32 | 49 | 1 | 1 | 14 | 80 | 31 | 30 | 0 | 11 | 6 | 7 | 0 | 0 | 23 | 5 | 22 | 0 | 2 | 23 | 32 | 19 | 25 | 0 | 4 |
| *Tephrocybe* sp. | 0 | 0 | 0 | 0 | 0 | 0 | 0 | 0 | 0 | 0 | 0 | 0 | 0 | 20 | 0 | 0 | 0 | 0 | 0 | 0 | 0 | 0 | 0 | 0 | 0 |
| *Thelephora terrestris* | 0 | 0 | 0 | 0 | 0 | 0 | 1 | 0 | 20 | 0 | 0 | 0 | 0 | 0 | 0 | 0 | 0 | 1 | 1 | 0 | 0 | 0 | 0 | 1 | 0 |
| *Tricholoma auratum* | 6 | 0 | 0 | 12 | 0 | 0 | 0 | 0 | 0 | 0 | 2 | 1 | 0 | 0 | 0 | 9 | 0 | 0 | 0 | 5 | 1 | 2 | 2 | 0 | 0 |
| *T. pessundatum* | 2 | 0 | 0 | 0 | 0 | 0 | 0 | 0 | 0 | 0 | 0 | 0 | 0 | 0 | 0 | 0 | 0 | 0 | 0 | 0 | 0 | 0 | 0 | 0 | 0 |
| *T. portentosum* | 4 | 0 | 0 | 0 | 0 | 0 | 0 | 0 | 0 | 0 | 0 | 0 | 0 | 0 | 0 | 0 | 0 | 0 | 0 | 0 | 3 | 0 | 0 | 0 | 0 |
| *T. saponaceum* | 1 | 0 | 0 | 0 | 0 | 0 | 0 | 0 | 0 | 0 | 0 | 0 | 0 | 0 | 3 | 0 | 0 | 0 | 0 | 8 | 0 | 0 | 0 | 0 | 0 |
| *Tricholomopsis rutilans* | 0 | 0 | 0 | 0 | 0 | 10 | 1 | 0 | 2 | 0 | 0 | 0 | 0 | 4 | 0 | 4 | 0 | 2 | 0 | 3 | 0 | 1 | 1 | 2 | 0 |
| *Tylopilus felleus* | 1 | 2 | 0 | 0 | 0 | 0 | 0 | 0 | 0 | 0 | 0 | 0 | 0 | 0 | 0 | 0 | 0 | 0 | 0 | 0 | 0 | 0 | 0 | 0 | 0 |
| *Xerocomus subtomentosus* | 0 | 0 | 0 | 0 | 0 | 0 | 0 | 0 | 0 | 0 | 0 | 1 | 0 | 0 | 0 | 0 | 0 | 1 | 0 | 0 | 0 | 0 | 1 | 1 | 0 |
| *Xeromphalina campanella* | 0 | 0 | 0 | 0 | 20 | 3 | 65 | 0 | 10 | 82 | 5 | 0 | 0 | 0 | 0 | 0 | 0 | 0 | 0 | 0 | 0 | 0 | 0 | 0 | 0 |
| | 503 | 300 | 113 | 371 | 280 | 299 | 190 | 100 | 52 | 664 | 123 | 532 | 201 | 72 | 422 | 318 | 172 | 26 | 168 | 375 | 258 | 268 | 95 | 90 | 144 |

There were 6 over-aged, 4 thinning stage, 8 pole stage and 7 sapling stage stands.

The amount of macrofungi fruit bodies was recorded by a line survey method. A 20–25 m broad zone along the borders of the stand was excluded and the rest of the stand was crossed by a 3-m-broad transect, from which all basidiocarps of macrofungi were collected, counted and partly weighed. The length of the transect in each stand was 250 m, which was measured by pacing. The transect was kept as straight as possible and two rows between planted trees were used alternatively. In each stand mushrooms were collected 6–11 times (Table 1). The number of collections was the same within the nine stand groups. In total, 11 542 fruit bodies were collected.

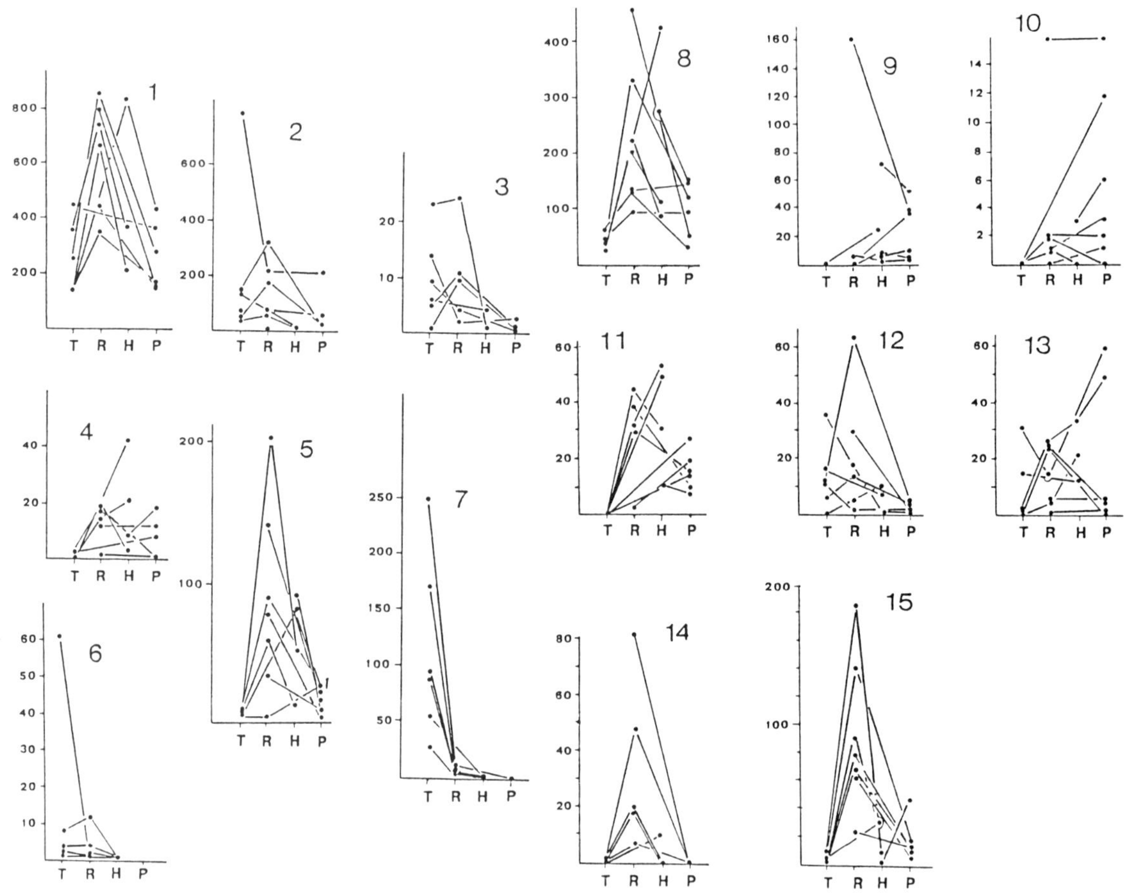


*Figs. 1–15.* The number of fruit bodies in stands of different development stages. Vertical axis: number of fruit bodies. Horizontal axis: T: sapling stands, R: pole stage stands, H: thinning stage stands and P: stands at final felling stage. The lines connect sample stands situated side by side. Fig. 1: total number of basidiocarps (lignicolous species excluded). 2: *Clitocybe* spp., 3: *Collybia* spp., 4: *Cantharellula umbonata*, 5: *Cystoderma amianthinum* coll., 6: *Pholiota spumosa*, 7: *Laccaria laccata* coll., 8: *Lactarius rufus*, 9: *Russula decolorans*, 10: *R. paludosa*, 11: *R. vinosa*, 12: *Suillus bovinus*, 13: *S. variegatus*, 14: *Cortinarius semisanguineus* and 15: *Cortinarius* spp.

## RESULTS

The results of the line surveys are given in Table 1 and in Figs. 1–15. The emphasis was laid on soil-inhabiting mycorrhizal and saprophytic species, as lignicolous species cannot be expected to show a direct correlation with the tree age and as some of the lignicolous species persist in stumps for over 10 years.

About 80 fungus species were found in the stands, although some were determined only to genus (*Cortinarius, Clitocybe*). The majority of the species occurred so sparsely and infrequently that no correlation with the age of the stand could be established, and for this reason only more abundant species are considered here.

As expected, the number of fruit bodies was lowest in sapling stands (see Fig. 1), but rose rapidly in the pole and thinning stages, and clearly decreased in over-aged stands.

Soil saprophytes, the genera *Collybia* and *Clitocybe* and *Pholiota spumosa*, were very clearly concentrated in stands at the sapling stage, and were found only occasionally in the final felling stage (Figs. 2, 3 and 6). On the other hand, *Cantharellula umbonata* and *Cystoderma amianthinum* (coll.), which are probably saprophytes, favoured pole stage stands (Figs. 4 and 5).

Among mycorrhizal species, the most numerous was *Lactarius rufus*, which composed 29% of all fruit bodies collected. The species was encountered in stands of all stages, although it was clearly most abundant in pole stage stands (Fig. 8).

Large *Russula* species (*R. vinosa, R. decolorans, R. paludosa*) mostly seemed to favour over-aged stands (Figs. 9–11), although the difference was not statistically significant.

16

*Figs. 16–18.* Fig. 16: stand 7, young regeneration area, Fig. 17: stand 8, pole-sized stand, where the number of basidiocarps was highest. Fig. 18: over-aged stand 2.

*Laccaria laccata* (coll.), which is a mycorrhizal species (Brown & Sinclair 1981), was very clearly concentrated in the sapling stage, in which it was sometimes found in masses, due to disturbance of the soil in connection with planting, and was almost totally lacking in the stage of final felling.

Among boletes, *Suillus bovinus* clearly favoured sapling stands, while *S. variegatus*, like the russulas, was more abundant in older stands.

Collectively, the genus *Cortinarius* very clearly favoured pole stage stands (Fig. 15), but also occurred fairly abundantly in older stands. *C. semisanguineus* was concentrated still more clearly in pole stage stands (Fig. 14).

According to the present material, the first mycorrhizal fungus species in young pine sapling stands are *Suillus bovinus, Lactarius rufus, Chroogomphus rutilus* and *Laccaria laccata*. The last-named species seems to be the most specialized mycorrhizal species of young saplings, though more thorough investigations on its ecology seem to be necessary.

## DISCUSSION

The age of the trees is only one of many factors affecting the occurrence of fungus species and production of their fruit bodies. Another factor is disturbance of the soil in connection with planting, which favours mainly *Laccaria laccata* and also *Lactarius rufus*. The abundant occurrence of *Laccaria* can partly be explained by this factor. Otherwise the effect of man seemed to be the same in all stands.

The results indicate fairly clearly that the conditions in pole-aged stands (Fig. 17) give the best possibilities for the production of fruit bodies of mycorrhizal species. Compared with sapling and over-aged stands (Figs. 16 and 18), the dense tree stand in pole-aged forests gives favourable conditions for growth of mycelia, because the humus and moss layers stay moist longer than in more open stands. In addition, the microclimate is less extreme in regard to frosts. The frequency of *Cystoderma amianthinum* and *Cantharellula umbonata* seems to be connected with the thicker moss and litter layers in these stands. Otherwise, the concentration of saprophytic species in sapling stands can be explained by the great amount of needle and branch litter left on the soil after the final felling.

The amount of litter falling on the soil seemed to be greatest in pole-aged and older stands, as they were fairly dense. This favours the activity of mycelia indirectly by increasing the thickness of the humus and moss layers compared with those in the regeneration areas, where the moss layer is often poorly developed and the humus layer thinner. According to Hari et al. (1982), the amount of branch, root and needle biomass is greatest in a pine stand at the age of 20–30 years, which is also the stage at which the number of mycorrhizal species is greatest. The growth of trees is also most rapid at this stage.

However, the physiological condition of the trees may affect the occurrence of mycorrhizal species. It is possible that small saplings do not have the same carbohydrate reserves as older trees. In over-aged trees, the products of photosynthesis may be needed for the maintenance of living tissues in stems and roots. According to a model presented by Hari et al. (1982), the net photosynthesis of a pine forest is greatest at the age of 20–30 years (as kg/ha/year).

According to investigations by Fleming (1983, 1984), Fleming et al. (1984) and Last et al. (1985), there are fungus species, which readily colonize the roots of birch seedlings at the first stages (*Paxillus involutus, Hebeloma* sp.). Other fungus species, such as *Lactarius pubescens*, colonize roots of seedlings later and especially if they are in contact with roots of older trees. Factors of this kind evidently also affect the occurrence of mycorrhizal species in stands of different ages.

The results suggest that both external (moisture, thickness of the humus, microclimate) and internal factors (amount of net photosynthesis) favour the growth of mycorrhizal fungi in pole-aged stands.

The results clearly suggest that the amount of edible mushrooms is greatest in relatively dense pole-aged stands, although russulas and *Suillus variegatus* may occur more abundantly in over-aged stands. The material for the most valuable commercial mushrooms, *Boletus edulis* and *Cantharellus cibarius* is so small that a separate study concentrating on these species will be necessary.

## REFERENCES

Brown, A.C. & Sinclair, W.A. 1981: Colonization and infection of primary roots of Douglas-fir seedlings by the ectomycorrhizal fungus Laccaria laccata. — Forest Sci. 27:111–124.

Fleming, L.V. 1983: Succession of mycorrhizal fungi on birch; infection of seedlings planted around mature trees. — Plant and Soil 71:263–267.

Fleming, L.V. 1984: Effects of soil trenching and coring on the formation of ectomycorrhizas on birch seedlings grown around mature trees. — New Phytol. 98:143–153.

Fleming, L.V., Deacon, J.W., Last, F.T. & Donaldson, S.J. 1984: Influence of propagating soil on the mycorrhizal succession of birch (Betula pendula) seedlings transplanted to a field site. — Transact. British Mycol. Soc. 82:707–712.

Hari, P., Kellomäki, S., Mäkelä, A., Ilonen, P., Kanninen, M., Korpilahti, E. & Nygren, M. 1982: Metsikön varhaiskehityksen dynamiikkaa. (Summary: Dynamics of early development of tree stand.) — Acta Forestalia Fennica 177:1–41.

Jäppinen, J.-P., Hotanen, J.-P. & Salo, K. 1986: Marja- ja sienisadot ja niiden suhde metsikkötunnuksiin mustikka- ja puolukkatyypin kankailla Ilomantsissa vuosina 1982–1984. Ref. Yields of wild berries and larger fungi and their relationship to stand chracteristics on MT and VT-type mineral soil sites in Ilomantsi, eastern Finland, 1982–1984. — Folia Forestalia 670. 25 pp. Helsinki.

Last, F.T., Mason, P.A., Wilson, J., Ingleby, K., Munro, R.C., Fleming, L.V. & Deacon, J.W. 1985: "Epidemiology" of sheathing (ecto)-mycorrhizas in unsterile soils: a case study of Betula pendula. — Proc. Royal Soc. of Edinburgh, B 85:299–315.

Mason, P.A., Wilson, J. & Last, F.T. 1983: The concept of succession in relation to the spread of sheathing mycorrhizal fungi on inoculated tree seedlings growing in unsterile soils. — Plant and Soil 71:247–256.

Rautavaara, T. 1947: Suomen sienisato. — 534 pp. Werner Söderström Oy. Porvoo.

Received 18.V.1987

Acta Bot. Fennica 136:95–98, 1988

# Fungal productivity of pine heaths in North-West Estonia

K. KALAMEES and S. SILVER

Kalamees, K. & Silver, S. 1988: Fungal productivity of pine heaths in North-West Estonia. — In: Vänninen, I. & Raatikainen, M. (eds.), Proceedings of the Finnish-Soviet symposium on timber forest resources in Jyväskylä, Finland, 25–29 August 1986: 95–98. Acta Bot. Fennica 136. Helsinki. ISSN 0001-5369 ISBN 951-9469-30-3

The paper gives an account of observations made during four years (1978–1981) on the fungal productivity of two types of pine heaths (*Pinetum cladoniosum* and *P. callunosum*) in North-West Estonia.

Key words: Estonia, fungal productivity, pine heaths

*K. Kalamees, Institute of Zoology and Botany, Academy of Sciences of the Estonian SSR, 21 Vanemuise St., SU-202400 Tartu, Estonian SSR, USSR; S. Silver, Ministry of Forest Management and Nature Conservation of the Estonian SSR, Nova Forest District, SU-203175 Nova PO, Estonian SSR, USSR*

## MATERIAL AND METHODS

Investigations were carried out by the authors in the Nova and Vihterpalu Forest Districts, in the years 1978–1981. The productivity of edible fungi in the pine heaths of North Estonia has been studied earlier by Lougas (1981) in Lahemaa National Park.

According to the classification of forest site types in use in Estonia at present (Lõhmus 1984), pine heaths comprise two kinds of forest site types: the *Cladonia* type and the *Calluna* type. These are the driest forest site types occurring on the sandy soils of Estonia, being found mainly in North Estonia, along the coast of the Gulf of Finland, on the islands and, to a lesser extent, in South-East and South-West Estonia.

We carried out the observations in both forest site types, with *Pinus sylvestris* dominating in the tree layer. Seven permanent sample plots were used, of which three were situated in an 80-year pine heath of *Cladonia* site type, three in a 100-year pine heath of *Calluna* site type and one in 25-year pine wood of *Cladonia* site type. All the sample plots were covered with V–Va bonity class tree stands with an average closure of 0.7. In the fully grown forest, the mean height of the trees was 14 m and the trunk diameter 18 cm, in the young pine wood the corresponding measurements were 3 m and 6 cm. The forest types were recognized by their characteristic dwarf shrub and moss layers. The choice of permanent sample plots was based on the differences in the coverages of those layers in different parts of a forest site type. The dominant species in the dwarf shrub layer were *Calluna vulgaris* (L.) Hull, *Vaccinium vitis-idaea* L. and *Empetrum nigrum* L., while in the moss layer the most abundant species were *Cladonia* spp. and *Cetraria islandica* (L.) Ach. In both forest types the grass layer was represented by only a few plants. In the shrub layer there were solitary junipers (*Juniperus communis* L.), some alders (*Alnus incana* (L.) Moench) in two sample plots, and birches (*Betula pendula* Roth). The presence of those two tree species enriched the species composition of the mycorrhizal fungi in the area investigated. Both the *Cladonia* and *Calluna* types had arid podzol soils of high acidity.

The sample plots were quadratic or rectangular, with a size of 0.1 ha. During the growing season (from April to November), all the fruit bodies of edible macrofungi were harvested from the sample plots, the wormy specimens being separated from the unspoilt ones. All the fruit bodies were counted and weighed while fresh by species. To obtain an idea of the total fungal production and to calculate the ratio of edible to inedible fungi, all the inedible and poisonous fungi were also recorded. Their fruit bodies were also counted and weighed while fresh, but in most cases not by species, only by genera. For that reason data are not available on the total number of fungal species or the number of inedible species in the area. The fruit bodies were gathered at average intervals of 10 days.

## RESULTS AND DISCUSSION

Of the total 31 species of edible fungi recorded, 23 were growing in the 80-year-old pine heath of *Cladonia* type site, 11 in the young pine heath of *Cladonia* type and 24 in the pine heath of *Calluna* type (Table 1).

The most widespread species of inedible and poisonous fungi on the sample plots were *Dermocybe semisanguinea* (Fr.) Gill., *D. cinnamomea* (L.: Fr.) Wunsche s.l. and *Paxillus involutus* (Batsch: Fr.) Fr., which grew in all three areas. The two mature forest types had fairly frequent occurrences of

*Lactarius helvus* (Fr.: Fr.) Fr., *Collybia peronata* (Bolt.: Fr.) Kumm. and *Amanita porphyria* (Alb. & Schw.: Fr.) Secr. The species of the genera *Dermocybe* and *Cortinarius* were fairly numerous in all the forest types, being the most productive among the inedible fungi.

The average numbers of fruit bodies and the yields in the forest site types studied are presented in Tables 2 and 3.

As can be concluded from Tables 2 and 3, both the number of fruit bodies and the production in kilograms per hectare vary fairly widely from year to year within one forest type. In the mature forests, the differences between the highest (1980) and the lowest (1979) yields of edible fungi were as great as 2–3-fold (Table 3).

In the young pine wood the yields were more constant, the lowest being that of 1981. The production of inedible fungi in mature forests was in most cases only 1/2–1/4 of that of edible fungi, but the numbers of inedible fungi were generally twice as great as those of edible fungi. In all the study years the young wood harboured relatively few inedible fungi. The worminess of the fruit bodies was rather high in mature woods: judging by the number of fruit bodies, almost 1/3–1/2 of the yield (Table 3); when calculated as the weight of the fruit bodies, the part of the yield that could not be used could be as much as 1/3. In the young pine wood, the worminess percentage was somewhat lower (up to 20%).

According to the average production of the two mature forest types in the years studied (Table 3), the *Calluna* forest type was more productive: the total fungal yield being 199 kg/ha, and that of edible fungi 143 kg/ha. The corresponding values for the *Cladonia* site type were 188 kg/ha and 124 kg/ha. The average percentage of worminess was high (66%) in the *Calluna* site type, but considerably lower in the *Cladonia* type (51%). Thus the unspoilt yield of edible fungi in the *Cladonia* site type was higher (64 kg/ha) than in the *Calluna* site type (47 kg/ha). The number of fruit bodies of edible fungi was almost the same in the two mature forest sites (Table 2).

The fungal production of the young pine wood was extremely high; the total yield was 504 kg/ha, edible fungi contributing 499 kg/ha and inedible fungi only 5 kg/ha (Table 3). Due to the small proportion of wormy fungi (11%), the un-

Table 1. *The species composition of edible fungi in the pine heaths of North-West Estonia in 1978–1981 (+ denotes the presence of a species; they are arranged in order of the years).*

| | *Pinetum cladoniosum* 80 years old | *Pinetum cladoniosum* 25 years old | *Pinetum callunosum* 100 years old |
|---|---|---|---|
| *Armillaria mellea* (Vahl: Fr.) Kumm. s.l. | − − − − | − − − − | + − − − |
| *Chroogomphus rutilus* (Schaeff.: Fr.) Miller | + + + − | + + + − | + + + + |
| *Cortinarius collinitus* (Sow.: Fr.) Fr. | + + + − | − + − − | + + − − |
| *C. mucosus* (Bull.: Fr.) Kickx | + + + + | − + − − | + + + + |
| *Gomphidius roseus* (Fr.) P. Karst. | + + + − | + + + + | − − − − |
| *Hygrophoropsis aurantiaca* (Wulf.: Fr.) Maire | − − − − | − − − − | − − + − |
| *Hygrophorus hypothejus* (Fr.: Fr.) Fr. | + − − − | + + + + | − − − − |
| *Hypholoma capnoides* (Fr.: Fr.) Kumm. | − − − − | − − − − | + + + − |
| *Laccaria laccata* (Scop.: Fr.) Berk. & Br. s.l. | − + + + | − + + + | − − + + |
| *Lactarius lilacinus* (Lasch) Fr. | + − − − | − − − − | − − − − |
| *L. rufus* (Scop.: Fr.) Fr. | + + + + | + + + + | + + + + |
| *L. torminosus* (Schaeff.: Fr.) S.F. Gray | − − − − | − − − − | − − − + |
| *Leccinum scabrum* (Bull.: Fr.) S.F. Gray | − − − − | − − − − | − + − − |
| *L. testaceoscabrum* (Secr.) Sing | − − − − | − − − − | − + − − |
| *L. vulpinum* Watl. | − − − − | − − − − | − + + − |
| *Rozites caperata* (Pers.: Fr.) P. Karst. | + + + + | − − − − | + + + + |
| *Russula adusta* (Pers.: Fr.) Fr. | + + + + | − − − − | − + − − |
| *R. decolorans* (Fr.) Fr. | − + − − | − − − − | + + + + |
| *R. emetica* (Schaeff.: Fr.) S.F. Gray s.l. | − + + + | − − − − | − + + + |
| *R. paludosa* Britz. | + + + + | − − − − | + + + + |
| *R. vinosa* Lindbl. | − − − + | − − − − | − − − − |
| *Sarcodon imbricatus* (L.: Fr.) P. Karst. | − + + − | − − − − | − − − − |
| *Suillus bovinus* (L.: Fr.) Kuntze | + + + + | + + + + | + + + + |
| *S. granulatus* (L.: Fr.) Kuntze | + − − − | − + − − | − + − − |
| *S. luteus* (L.: Fr.) S.F. Gray | + − − − | + − − − | − − − − |
| *S. variegatus* (Sow.: Fr.) Kuntze | + + + + | + − + + | + + + + |
| *Tricholoma flavoviridens* (Pers.: Fr.) Lund | + + + + | − − − − | − + + − |
| *T. portentosum* (Fr.) Quél. | − − − − | − − − − | − − + − |
| *Tricholomopsis rutilans* (Scaeff.: Fr.) Sing. | + − + − | − − − − | + + + + |
| *Xerocomus badius* (Fr.) Kuhner & Gilb. | − − + − | − − − − | − − − − |
| *X. subtomentosus* (L.: Fr.) Quél. | + + + + | − − − − | − + − − |

Table 2. *The average number of fungal bodies per hectare in the pine heaths of North-West Estonia in 1978–1981 (accuracy ± 1 fruit body per ha).*

| | Year | Total no. of fungi | Edible fungi | | | | Inedible fungi |
|---|---|---|---|---|---|---|---|
| | | | Total no. | Unspoilt | Wormy | % of worminess | |
| *Pinetum cladoniosum* | 1978 | 10 753 | 5 903 | 3 673 | 2 503 | 43 | 4 850 |
| 80 years old | 1979 | 16 687 | 5 670 | 3 607 | 2 063 | 36 | 11 017 |
| | 1980 | 25 443 | 9 960 | 7 007 | 2 953 | 30 | 15 483 |
| | 1981 | 14 853 | 4 570 | 2 423 | 2 147 | 47 | 10 283 |
| | Average | 16 934 | 6 522 | 4 173 | 2 423 | 39 | 10 412 |
| *Pinetum cladoniosum* | 1978 | 55 050 | 54 840 | 49 140 | 5 700 | 11 | 210 |
| 25 years old | 1979 | 48 890 | 47 200 | 45 470 | 1 730 | 4 | 1 690 |
| | 1980 | 58 260 | 55 990 | 53 010 | 2 980 | 5 | 2 270 |
| | 1981 | 40 720 | 40 110 | 33 270 | 6 840 | 17 | 610 |
| | Average | 50 730 | 49 535 | 45 223 | 4 313 | 9 | 1 195 |
| *Pinetum callunosum* | 1978 | 18 273 | 6 440 | 2 530 | 3 910 | 61 | 11 833 |
| 100 years old | 1979 | 11 123 | 3 733 | 1 970 | 1 763 | 47 | 7 390 |
| | 1980 | 20 760 | 9 780 | 5 220 | 4 560 | 47 | 10 980 |
| | 1981 | 10 737 | 6 013 | 2 587 | 3 427 | 57 | 4 723 |
| | Average | 15 223 | 6 492 | 3 077 | 3 415 | 53 | 8 732 |

spoilt yield was high, 444 kg/ha. The number of fruit bodies was also very high, totalling 50 730/ha, of which 49 535 were edible fungi.

In the 80-year-old pine heath of *Cladonia* site type, the most productive species of edible fungi were *Suillus variegatus, Rozites caperata* and *Lactarius rufus*. Their average yields were: *Suillus variegatus* 41 kg/ha, *Rozites caperata* 33 kg/ha and *Lactarius rufus* 20 kg/ha. *Rozites caperata* gave the maximum yield in 1980 (61 kg/ha). The average number of fruit bodies was highest in *Lactarius rufus* — 1 755/ha. The highest absolute number of fruit bodies was also achieved by this species, 3 643/ha in 1980.

In the 100-year-old pine heath of *Calluna* site type, the following species of edible fungi were the most productive: *Russula decolorans* 53 kg/ha, *R. paludosa* 35 kg/ha and *Lactarius rufus* 24 kg/ha. The productivity of *Rozites caperata* was especially low in the *Calluna* type, averaging 1.7 kg/ha. Areas with a thick dwarf shrub layer did not seem to be favourable for this species. *Russula decolorans* gave the highest yield, 94 kg/ha (in 1980). In this site type also, the greatest average

Table 3. The average production of fungal fruit bodies (kg/ha) in the pine heaths of North-West Estonia in 1978–1981 (accuracy ± 1 kg per ha).

| | Year | Total yield of fungi | Edible fungi | | | | Inedible fungi |
|---|---|---|---|---|---|---|---|
| | | | Total yield | Unspoilt | Wormy | % of worminess | |
| *Pinetum cladoniosum* | 1978 | 143 | 105 | 52 | 53 | 50 | 37 |
| 80 years old | 1979 | 138 | 85 | 41 | 44 | 52 | 53 |
| | 1980 | 305 | 215 | 126 | 89 | 41 | 91 |
| | 1981 | 139 | 92 | 37 | 55 | 60 | 47 |
| | Average | 188 | 124 | 64 | 60 | 51 | 57 |
| *Pinetum cladoniosum* | 1978 | 549 | 548 | 477 | 72 | 13 | 1 |
| 25 years old | 1979 | 445 | 440 | 424 | 16 | 4 | 5 |
| | 1980 | 581 | 569 | 529 | 40 | 7 | 12 |
| | 1981 | 440 | 438 | 345 | 93 | 20 | 3 |
| | Average | 504 | 499 | 444 | 55 | 11 | 5 |
| *Pinetum callunosum* | 1978 | 183 | 97 | 36 | 61 | 63 | 85 |
| 100 years old | 1979 | 122 | 85 | 33 | 51 | 61 | 37 |
| | 1980 | 299 | 239 | 71 | 168 | 70 | 60 |
| | 1981 | 192 | 152 | 46 | 106 | 70 | 40 |
| | Average | 199 | 143 | 47 | 97 | 66 | 56 |

number of fruit bodies was recorded for *Lactarius rufus* — 1 793/ha. The highest absolute number, 3 323 fruit bodies per ha, was recorded for *L. rufus* in 1980.

In the young pine wood of *Cladonia* site type, the yield of edible fungi was mainly contributed by two species: *Lactarius rufus* and *Suillus bovinus*. The former was very productive — 405 kg/ha, while the latter yielded 100 kg/ha. The number of their fruit bodies was also very high: *Lactarius rufus* averaged 40 585/ha and *Suillus bovinus* 10 288/ha. In 1980 as many as 50 070 fruit bodies of *Lactarius rufus* were gathered per hectare.

Up to 70% of the yields of these species was worm-eaten, but in the case of *L. rufus* the percentage was very low, averaging three in all the forest types studied. Consequently, almost all the natural yield of *L. rufus* can be used, while in the other species only 1/3–1/2 of the total yield can be eaten.

In the pine heath of *Cladonia* site type, the fruit bodies of the species mentioned above have proved to be most abundant from mid-August to mid-October. This period may thus be considered the peak of the summer–autumn fungal season in the pine heaths of North-West Estonia.

In the pine heath of *Calluna* site type, the dynamics of the more productive edible fungi differed from that in the *Cladonia* site type. In the *Calluna* type those species ordinarily had two peaks, one in the last decade of July, or the first decade of August and the other at the end of August or in September. As regards the number of fruit bodies, these peaks were often equal, but still more often one of the peaks exceeded the other peak generally being the greatest.

Comparison of our data with those obtained by Lougas (1980) shows that during 2 months (1.IX.–2.XI.1975) an 80-year-old *Cladonia* heath in Lahemaa National Park gave a considerably larger yield of edible fungi, 204.7 kg/ha. At the same time, a greater proportion of the fungi was worm-eaten (62%) than in our study area, so that the unspoilt yield was only 78 kg/ha. This is not much higher than the unspoilt yield gathered by us from the pine heath of *Cladonia* site type (Table 3). The total crop of edible fungi reported by Lougas from pure pine heath of the *Calluna* site type was 50 kg/ha, which is a third of that obtained in our study area.

The production of edible fungi recorded by Rautavaara (1947) in a pine heath of the *Cladonia* site type in Finland was 82.5 kg/ha. Ohenoja (1980) reported that the average production of edible fungi in *Cladonia* and *Calluna* site type forests during good fungus years (1977, 1978) in Finland was nearly 80 kg/ha. In the Kirov region, Skrjabina and Larina (1967) reported a total yield of 106 kg/ha, the unspoilt production being 57.2 kg/ha.

## REFERENCES

Lõhmus, E. 1984: Eesti metsakasvukohatüübid. — 88 pp. Tallinn.

Lougas, T. 1980: Söögiseente varudest mõnedes Lahemaa Rahvuspargi metsakasvukohatüüpides. — Eesti Looduseuurijate Seltsi Aastaraamat 68:32–49.

Ohenoja, E. 1980: Sienisatotutkimus vv. 1976–78. — EKT-series 548:1–42. Inst. Food Chemistry Technol., Univ. Helsinki.

Rautavaara, T. 1947: Suomen sienisato. — 534 pp. Porvoo & Helsinki.

Skrjabina, A.A. & Larina, T.G. 1967: Urozhainost' s'edobnyh gribov v raznykh tipakh lesa. — Rastit. resursy 3:202–204.

Received 28.VII.1987

Acta Bot. Fennica 136:99–103, 1988

# Insect infestation of edible mushrooms in Soviet South Karelia and bioecological characteristics of the pests

E. B. YAKOVLEV

Yakovlev, E.B. 1988: Insect infestation of edible mushrooms in Soviet South Karelia and bioecological characteristics of the pests. — In: Vänninen, I. & Raatikainen, M. (eds.), Proceedings of the Finnish-Soviet symposium on timber forest resources in Jyväskylä, Finland, 25–29 August 1986: 99–103. Acta Bot. Fennica 136. Helsinki. ISSN 0001-5369 ISBN 951-9469-30-3

The production of fruit bodies and their infestation by insects were studied in the forests surrounding Petrozavodsk since 1977. In pine and pine–spruce stands, the most frequent and productive edible species were *Leccinum* spp., *Russula* spp., *Suillus variegatus, Lactarius rufus* and *Rozites caperatus*; in birch forests — *Russula* spp. and *Lactarius* spp., in aspen forests — *Leccinum* spp. The *Boletus edulis* group, *Suillus luteus* and *Russula* spp. have the highest degree of infestation (80–90%). The main groups of pests are the larvae of the dipterous families Mycetophilidae, Anthomyiidae and Phoridae. The infestation of Leccinum species averaged 50–60%, their principal pests being Anthomyiidae. The same percentage of infestation was found for *Suillus variegatus, Rozites caperatus, Lactarius necator, L. torminosus* and *Verpa bohemica*, but here the main pests were Mycetophilidae. The biological data on the insects injuring young and old sporophores, and some characters of the consortive links between groups of "lower" and "higher" Diptera and different groups of larger fungi are briefly discussed.

Key words: Diptera, ecology, edible fungi

*E.B. Yakovlev, Forest Research Institute of the Karelian Branch of the USSR Academy of Sciences, Pushkinskaya 11, SU-185610 Petrozavodsk, USSR*

Study of the species composition and bioecological characteristics of insects living in mushroom fruit bodies was begun in Soviet South Karelia in 1977. The faunistic investigations cover the surroundings of Petrozavodsk within a radius of 100 km. 210 species of dipterous insects from about 90 species of the fungus orders Agaricales s.l., Aphyllophorales and Pezizales were reared, and also 29 beetle species were obtained from the examined sporophores. Many of these insect species had already been observed to be mycetobionts in Finland (Hackman & Meinander 1979). The bulk of the material was obtained from edible mushrooms. Figs. 1–3 show the main groups and principal species of insects inhabiting the most valuable mushrooms picked in the area.

The fruit body production of the larger fungi and the insects inhabiting them were examined in permanent plots in forests of different site types in the Kivach State Nature Reserve in 1981, 1983 and 1984. Table 1 gives some data on the infestation of *Leccinum scabrum, Suillus bovinus, S. luteus, Lactarius necator* and *L. torminosus* in 1981, of the *Boletus edulis* group and *Rozites caperatus* in 1983 and 1984, and of the *Leccinum versipelle* group, *Suillus variegatus* and *Russula* spp. in the three years. More complete results have been published elsewhere (e.g. Yakovlev & Osipova 1985, Krivosheina et al. 1986, Yakovlev 1986a, b).

## INSECT FAUNAS OF DIFFERENT MUSHROOM GROUPS

The mushrooms traditionally gathered in Karelia mainly belong to the family Boletaceae. The species of the *Boletus edulis* group are considered the most valuable. *B. pinophilus* fruits abundantly in pine stands (*Pinus sylvestris*), especially in young forest of the Cladonia and Calluna site types, but its use is limited by its great liability to damage by insects. Insect infestation is lower only in those years when the mass fruiting takes place in September. The yields of *Boletus* are mainly damaged by the larvae of the fungus gnats (Mycetophilidae) and anthomyiid flies (Anthomyiidae), the frequency of which varies from one year to another. The most common pest species are *Mycetophila fungorum* (De Geer), *Cordyla brevicornis* Staeg., *Exechia separata* Lundstr. and *Pegomya tenera* (Zett.). In some years the fruit bodies of *Boletus* species are often inhabited by the larvae of such flies as *Megaselia rubella* Schmitz, *M. rubescens* Wood. (Phoridae) and *Tephrochlamys flavipes* (Zett.) (Heleomyzidae).

In the mushroom genus *Suillus*, the most productive in upland pine stands and spruce (*Picea abies*) stands with pine is *S. variegatus*. On average, about half of all the fruit bodies examined in this fungus had been damaged mainly by larvae of

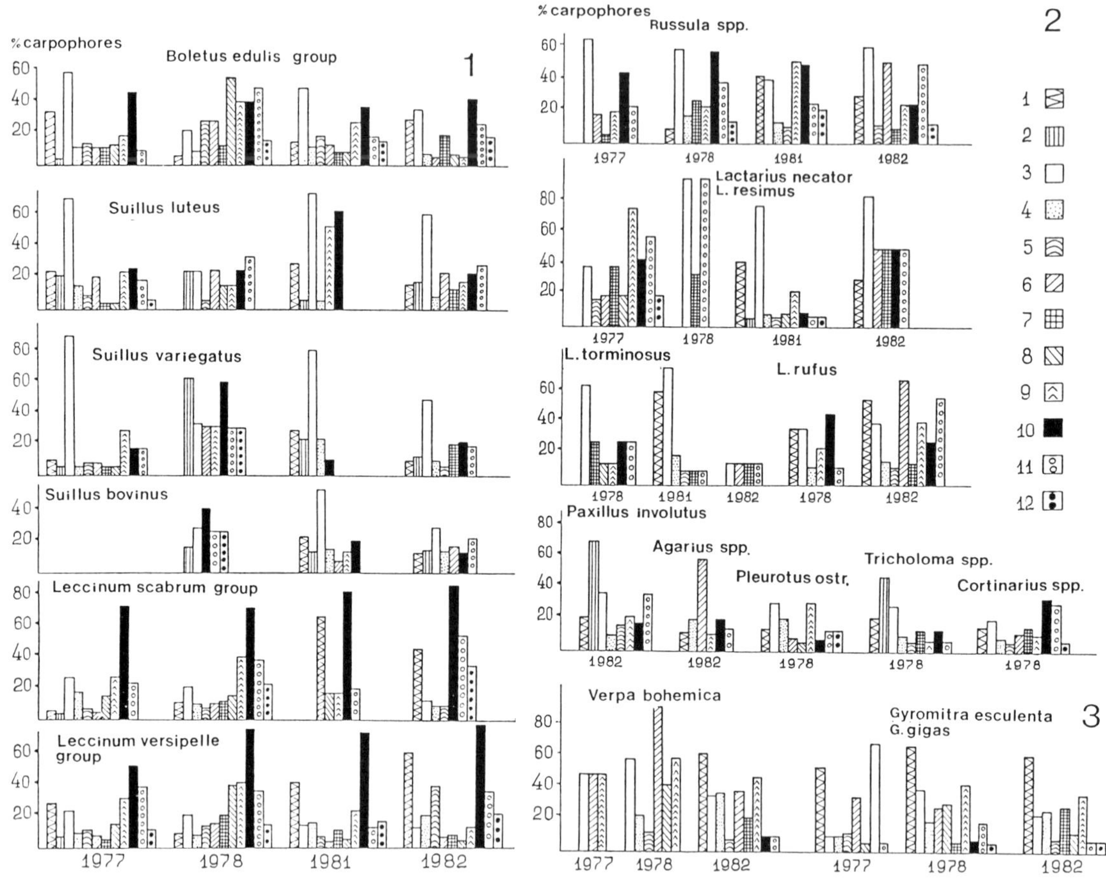


*Figs. 1–3.* — 1 (left): The main groups of dipterous insects inhabiting Boletaceae. Symbols as in Fig. 2. 2 (above right): The main groups of dipterous insects inhabiting Agaricales and Russulales. 1: Limoniidae, 2 = Bolitophilidae, 3 = Mycetophilidae, 4 = Sciaridae, 5 = Cecidomyidae, 6 = Phoridae, 7 = Heleomyzidae, 8 = Sphaeroceridae, 9 = Drosophilidae, 10 = Anthomyiidae, 11 = Muscidae, 12 = Fanniidae. 3 (below right): The main groups of dipterous insects inhabiting Pezizales. Symbols as in Fig. 2.

mycetophilids: *Mycetophila fungorum, Exechia separata* and *Exechiopsis indecisa* (Walk.), among others, and also *Bolitophila rossica* Landr. (Bolitophilidae) in autumn. The damage was more serious in *Suillus luteus* and *S. bovinus*; besides the same species of Mycetophilidae, they commonly harboured the following flies: *Pegomya deprimata* (Zett.) (predominant), *P. tenera, P. haemorrhoa* Zett. (less common), *Megaselia rubella* and *M. rubescens*.

Other prized mushrooms are the species of *Leccinum*, fruiting abundantly in deciduous and coniferous forests. The proportion of damaged fruit bodies ranged from 2/3 (*L. versipelle* group) to one half (*L. scabrum* group). The main pests are anthomyiids: *Pegomya incisiva* Stein, *P. tenera, P. transversa* (Fall.), *P. pilosa* Stein, among others. Mycetophilidae are rather sparse and are mainly represented by other species than those found in *Boletus* and *Suillus*. The most common species are *Mycetophila signatoides* Dz. and *Rondaniella dimidiata* (Mg.)

In all the species of Boletaceae examined, 20–30 per cent of the mushrooms were inhabited in each year by polyphagous species, among which Limoniidae (*Ula sylvatica* (Mg.), less common; *Metalimnobia bifaciata* (Schrank) and Drosophilidae (*Drosophila transversa* Fall., less common; *D. busckii* Coq.; *D. funebris* (Fabr.) predominated. Many dipterous species of Sciaridae, Cecidomyiidae and Heleomyzidae, and one syrphid fly, *Cheilosia scutellata* (Fall.), are also very common in the Boletaceae, particularly *Leccinum*. Caps of *Leccinum* are often eaten through by staphylinid beetles of the genus *Oxyporus: O. rufus* (L.), *O. maxillosus* F., *O. mannerheimi* Gyll, and their stipes by larvae of Elateridae (wireworms).

Mushroom species of the family Gomphidiaceae, ecologically close to Boletaceae, also resemble these in their pests. Larvae of *Megaselia rubella, Pegomya haemorrhoa, P. transversa* and *P. deprimata* are most common. Among mycetophilids, the typical inhabitants of caps, *Exechia separata* and *Cordyla brevicornis* were noted here. The same can be said of

Table 1. *The fruit body production of some edible mushrooms and their infestation by insects in forests of different site types in the Kivach State Reserve (1981, 1983 and 1984).*

| | Fruit body biomass (kg/ha) | | | | Infestation, % | | | |
|---|---|---|---|---|---|---|---|---|
| | Pineta[1] | Piceeta–Pineta[2] | Betuleta[3] | Tremuleta[4] | Pineta[1] | Piceeta–Pineta[2] | Betuleta[3] | Tremuleta[4] |
| *Boletus edulis* group | 81.4 | – | 3.0 | 6.3 | 94 | – | 100 | 60 |
| *Leccinum versipelle* group | 93.6 | 41.1 | – | 87.1 | 59 | 73 | – | 72 |
| *L. scabrum* group | 11.0 | – | 8.6 | – | 46 | – | 42 | – |
| *Suillus variegatus* | 87.7 | 72.9 | – | – | 52 | 48 | – | – |
| *S. bovinus* | 23.7 | – | – | – | 78 | – | – | – |
| *Lactarius necator* | 3.1 | 21.0 | 27.3 | – | 100 | 12 | 40 | – |
| *L. torminosus* | – | 2.4 | 25.5 | – | – | 69 | 53 | – |
| *L. rufus* | 40.7 | 31.4 | 22.2 | – | 16 | 42 | 9 | – |
| *Russula* spp. | 64.6 | 139.3 | 56.0 | 8.3 | 85 | 83 | 62 | 87 |
| *Rozites caperatus* | 48.1 | 5.8 | 13.1 | – | 76 | 67 | 26 | – |

1) Pineta hylocomioso and Pineta cladinosa groups
2) Piceeta–pineta hylocomioso–cladinosa saxatilia group
3) Betuletum myrtilloso-mixto-herbosum
4) Tremuletum graminoso-mixto-herbosum

*Rozites caperatus*, but the species composition of its pests is much richer and infestation is higher than in the Gomphidiaceae.

*Paxillus involutus* yields are damaged by no more than 10–15 per cent (Shubin & Yakovlev 1983). This mushroom harbours only a small number of insect species, which may be due to the special biochemical composition of its fruit bodies. The larvae of bolitophilid gnats: *Bolitophila hybrida* and *B. nigrolineata* Landr. predominate.

The species of the genus *Agaricus*, which can be found in city parks and forests with rich soils, are usually damaged by phorid fly larvae, mostly *Megaselia rubella* and *M. rubescens*. Some of the carpophores examined contained larvae of *Ula sylvatica*, *Drosophila* sp. and *Pegomya* sp. No Mycetophilidae were found here. From the ephemeral fruit bodies of *Coprinus comatus*, also occurring in parks, *Drosophila transversa* and *D. funebris* were reared. Ephemeral habitats are rather typical of the larvae of these flies, but together with them some specimens of *Mycetophila fungorum* were obtained, whose larval development must undoubtedly last longer than the *Coprinus* fruit bodies.

One economically important group of edible mushrooms is the Russulaceae. The species of the genus *Russula* are more liable to be damaged by insects than those of *Lactarius* (milk-caps). Twenty-one species of Mycetophilidae were reared from the examined russulas; the principal pests were the following: *Mycetophila fungorum* (in one of every three of the sporophores examined), a number of species of the *Mycetophila ruficollis* group (mainly *M. strobli* Last.) and *Cordyla brevicornis*. The next most important groups of pests were the Anthomyiidae (mainly *Pegomya transversa* and *P. univittata* (v. Roser) and the Phoridae (e.g. *Megaselia rubella, M. rubescens*). The dominant pest group of milk-caps is also composed of mycetophilids (17 species, the most common are species of the *Mycetophila ruficollis* group and *Exechia nigroscutellata* Landr.). Anthomyiidae were not found so often as in Russula, and among them such species as *Pegomya tenera* and *Pegohylemyia silvatica* R.-D. predominated. Both russulas and milk-caps are often damaged by the larvae of *Ula sylvatica* (*L. torminosus* also by the larvae of *Metalimnobia bifasciata*) and certain flies such as *Drosophila transversa* and *Suillia* species: *S. atricornis* Mg., *S. bicolor* (Zett.), *S. flava* (Mg.). The gills of russulas are often eaten by staphylinid beetles of the genus *Gyrophaena*: e.g. *G. affinis* Mannerheim, *G. pulchella* Heer, *G. joyiodes* Wüsth.

Among the edible mushrooms growing on trees, the Karelians use mainly *Armillaria mellea*, which is sometimes damaged by larvae of Mycetophilidae, mainly *Mycetophila fungorum*. In late autumn, when its fruit bodies are not so badly infested, larvae of *Bolitophila cinerea* Mg. are found. In old sporophores larvae of *Ula sylvatica* and beetles belonging to the family Nitidulidae were found. The sporophores of *Pleurotus* are not so abundant in the wild of Karelia. They are often heavily infested and contain the larvae of such dipterous species as *Brachypeza armata* Winn., *B. hilaris* Winn., and *Drosophila trivittata* Strobl., which are never found in mushrooms growing in the ground. Old sporophores also harboured larvae of *Ula sylvatica* and imagos of the beetles *Triplax scutellaris* Charp., *Cyllodes ater* (Hbst.), *Mycetophagus quadripustulatus* (L.) and *Endomychus coccineus* (L.)

*Verpa bohemica*, occurring abundantly in very small plots in deciduous forests in spring time, is often infested by mycetophilid larvae, mostly species of *Allodia*: *A. neglecta* Edw., *A. sylvatica* Landr., *A. barbata* Lundstr., *A. ornaticollis* Mg. Some apothecia contain larvae of the flies *Drosophila transversa, Megaselia rubella, M. rubescens, M. mixta* (Schmitz), *Minilimosina fungicola* Hal., *Suillia nemorum* (Mg.), etc. Damage by insects is less serious in *Gyromitra esculenta*, occurring in great quantity in cut and burnt-over areas, and in the less abundant *Gyromitra gigas*. The only common inhabitants of these fungi are polyphagous dipterans: *Ula sylvatica, Drosophila transversa* and some species of the Phoridae, Cecidomyiidae, Sciaridae and Sphaeroceridae. Mycetophilidae are not very typical of *Gyromitra* (including the autumn-fruiting species); only the larvae of *Sciophila* species, *S. lutea* Macq., *S. kareliaca* A. Zaitzev and *S. modesta* A. Zaitzev, were found in *Gyromitra* more often than in other fungi, in the inner surface of the caps.

## ON ADAPTATIONS AND EVOLUTION OF INSECTS INFECTING MUSHROOMS

Thus the main destroyers of edible mushrooms are dipterous insects: gnats of the families Mycetophilidae and Bolitophilidae, and flies belonging to the genera *Pegomya* (Anthomyiidae), *Megaselia* (Phoridae) and *Suillia* (Heleomyzidae). These are the most specialized invaders of mushrooms, possessing certain bioecological adaptations. The chief of these is the decreased developmental period of the larvae (up to a week in summer), which corresponds to the period of ripening of the mushroom fruit bodies. The eggs of these insects can be found even in small, newly emerged carpophores. The larvae generally occur in fresh carpophores and leave the rotting ones for soil or litter. The migrating larvae appear to be full-fed, because they immediately begin to pupate. These insect species are characterized by host selection; they are commonly found in mushrooms which belong to one or a few genera. The species clearly preferring to occupy mushrooms of the same genus are of particular interest. Such specialist pest species are very characteristic of mushrooms growing on trees: *Lentinus, Pleurotus, Tricholomopsis, Hypholoma*, etc. A tendency towards monophagy was also noted in some insect species inhabiting such mushrooms as *Suillus, Leccinum, Tricholoma, Inocybe*.

Less specialized mycetophagous insects, feeding on both live and decaying fruit body tissues, are also responsible for damage to edible mushrooms. In some species of the Limoniidae, Sciaridae, Cecidomyiidae and Fanniidae, the period of larval development is much longer than the time required for the fruit bodies to ripen. Some flies, such as Drosophilidae and Sphaeroceridae, develop rapidly (4–5 days), but nevertheless were mostly found in carpophores that had begun to rot. These insects do not usually show host preference; larvae of the same species may be found in a wide variety of fungi, often in both Ascomycetes and Basidiomycetes. Besides dipterans, this group of mushroom-inhabiting insects includes many mycetophagous species of beetles, which can be found in both fresh and decaying fruit bodies.

Fungivorous insects that need fresh fruit body tissues, even if only at the beginning of their development (bio-mycetophages), form the basis of the complexes of mushroom-inhabiting insects. Besides them, inside rotting carpophores and on their surface, occur many dipteran species that appear to be saprophagous and zoophagous. Among the predators are certain muscid fly species of the genera *Mydaea, Alloeostylus* (and probably certain Staphylinidae and parasitic Hymenoptera species) obligately connected with mushrooms and also eurybiontic species (e.g. *Muscina* species). Some of the predators visit mushrooms periodically.

The use of fungal fruit bodies as food and habitat for the larvae is one of the principal lines of evolution in dipterous insects. The appearance and gradual development of specialized fungivorous groups can be traced in all the main stages of the historical development of the Diptera (Yakovlev 1986b). It may also be noted that those specialized fungivorous dipteran groups which, according to the paleoentomological data (Rohdendorf 1964), seem to be most archaic (Platypezidae, Bolitophilidae, some Sciophilinae) are mainly connected with Aphyllophorales and some Agaricales growing on trees. The typical inhabitants of almost all mushroom species of the Agaricales s.l. are the more advanced specialized genera of the Mycetophilidae, mainly belonging to the subfamily Mycetophilinae, together with evolutionarily younger groups of flies, such as Syrphidae, Heleomyzidae and Anthomyiidae, which apparently began to use fruit bodies of larger fungi simultaneously with angiospermous tissues. The higher flies generally predominate in the fleshy fruit bodies of the boletes, particularly in Leccinum and *Boletus* species. This allows speculation about the possibility that the Boletaceae are phylogenetically more advanced than the other Basidiomycetes, and the possibility that mycetophiloid gnats have been replaced by more advanced dipterous groups.

## THE ROLE OF FUNGIVOROUS INSECTS IN FOREST ECOSYSTEMS

Due to their great variety and large numbers, the fungivorous insects play an important role in forest ecosystems. They begin the first stage of utilization of the great fruit body biomass by breaking down the mushroom tissues and possibly carrying putrefactive micro-organisms into them, as fruit bodies undamaged by insects often dry up without marked rotting. Dipterous larvae migrating from decaying carpophores, disseminate mushroom spores close to the ends of plant roots. Adult dipterans, flying from one mushroom to another, can also carry spores. So the symbiosis of insects and fungi, based on trophic and topical relations on one hand, and phoresia on the other, is mutually beneficial. At the same time certain morphological characteristics of the fruit bodies, such as high stipe and special structure of the hymenophore, clearly indicate adaptation to anemochory in most species of hymenomycetes. Moreover, many dipterous species are able to destroy young fruit bodies before the spores ripen; some of them, e.g. *Cordyla brevicornis, C. fasciata* Mg., *Pegomya incisiva*, develop only in the stipe, without penetrating the hymenophore. These insects, which suppress growth and cause premature rotting of unripe sporophores, can be considered mushroom parasites. Particular interest attaches today to the study of the ecology of fungivorous insects in view of its connection with the task of protecting the edible mushroom yields against their insect pests.

## REFERENCES

Hackman, W. & Meinander, M. 1979: Diptera feeding as larvae on macrofungi in Finland. — Ann. Zool. Fennici 16:50–83.

Krivosheina, N.P., Zaitsev, A.I. & Yakovlev, E.B. 1986: Nasekomye — razrushiteli gribov v lesakh Evropeiskoi chasti SSSR. — 309 pp. Moscow.

Rohdendorf, B.B. 1964: Istoricheskoe razvitie dvukrylovykh nasekomykh. — Trudy Paleontol. Inst. Akad. Nauk SSSR 100:1–311.

Shubin, V.I. & Yakovlev, E.B. 1983: Povrezdaemost' gribov i zaselyayushchie ikh nasekomye. In: Rastitelnye resursy v svyazi s pobochnymi pol'zovaniyami v lesakh Karelii: 96–118. — Petrozavodsk.

Yakovlev, E.B. 1986a: Nasekomye-mitsebionty yuzhoi Karelii (ekologofaunisticheskii spisok). In: Fauna i ekologiya chlenistonogikh Karelii: 83–123. — Petrozavodsk.

Yakovlev, E.B. 1986b: Osobennosti konsortivnykh svyazei nasekomykh s makromitsetami (Some characters of consortive links between insects and macromycetes). — Mikol. i fitopatol. 20:186–192.

Yakovlev, E.B. & Osipova, L.T. 1985: Vidovoi sostav i bioekologicheskie osobennosti nasekomykh-obitatelei plodovykh tel s'edobnykh gribov v yuzhnoi Karelii. In: Nasekomye i fitopatogennye griby v lesnykh ekosistemakh: 4–71. — Petrozavodsk.

Received 28.VIII.1987